GARDEN
ALLIES

Published in 2021 by Timber Press, Inc.
The Haseltine Building
133 S.W. Second Avenue, Suite 450
Portland, Oregon 97204-3527
timberpress.com

Printed in China
Text and cover design by Vincent James

ISBN 978-1-64326-008-2
Catalog records for this book are available from the
Library of Congress and the British Library.

GARDEN ALLIES

The Insects, Birds & Other Animals

That Keep Your Garden Beautiful and Thriving

FRÉDÉRIQUE LAVOIPIERRE

ILLUSTRATIONS BY CRAIG LATKER

Timber Press
Portland, Oregon

To my father, Michèl Lavoipierre,

inquisitive scientist and renaissance man,

who taught me to love nature and scholarship,

with particular attention to the insects.

And

From Craig, to Richard,

for his infinite patience.

CONTENTS

PREFACE

I have always loved gardens and insects. As far back as I can remember, I was peering into the heart of our backyard roses, stirring up leaf litter, and climbing the apple tree to see what lived within its broad branches. One early yet vivid memory is of sitting on the edge of my sandbox, patiently watching a monarch butterfly emerge in slow motion from its gold-spangled, jade-green chrysalis. Underneath my grandmother's hydrangea, an ant battle between two opposing colonies held me enthralled for hours. Garden leaf beetles looked like shining jewels to my wondering eyes—and the holes they left in leaves, inconsequential. Insects and gardens were inextricable in my childhood world. As it turns out, they are inextricable in the world we all share.

When I began to garden seriously, I knew I did not want to use pesticides. I was horrified to find that a popular organic pest management book suggested using all kinds of insect-killing poisons. Instead, I studied my garden and watched shimmering hover flies lay their eggs among aphid colonies, after which the aphids slowly disappeared. I rejoiced at the "parsley worm" caterpillar that turned into a glorious swallowtail butterfly—who can eat that much parsley anyway? At the community college where I studied horticulture, I gave the "correct" answers in my pest management course, then went home to the pleasures of my backyard menagerie. Later, I applied my garden learning to

Bluebird

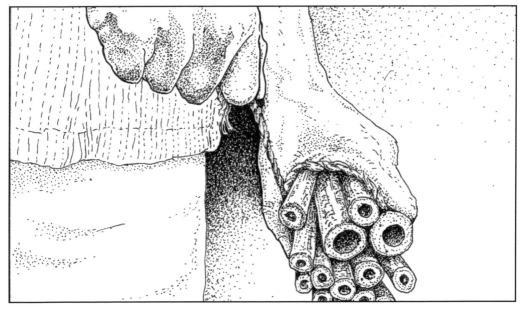

Clockwise from top left: Planting for beneficial insects; Home for bats; Bundling hollow reeds for bees

the nursery I owned. The natural enemies of pesky plant-munching insects became my friends and, fascinated, I later went on to learn more about them in my graduate studies.

My garden was beautiful, humming with life, and the nursery prospered. It helped that I was, and still am, an inherently lazy gardener. When my son was a baby, and I noticed an infestation of whiteflies on my currant bushes, I thought, "Oh, I really need to do something about this." But busy days went by and when I again visited that corner of the garden, the whiteflies were gone and I had a bounty of currants to enjoy! One day I noticed colonies of plump aphids jostling for space on a bank of lupines. Returning with my camera the next day, I discovered an army of soldier beetles feasting on them. Within a week, the aphids were gone and the soldier beetles had moved on.

Laziness, it turned out, paid off. So did curiosity. A close look at a mealy-bug colony revealed the camouflaged predator (a ladybug larva) in their midst. Solitary bees at a path's edge led me to leave a bit of bare soil for them to nest. My milkweed patches were a crowded natural-enemy nursery. A hand lens revealed that the hordes had descended—the plants were festooned with orange aphids! The banquet attracted hunting and parasitic wasps, and the predatory larvae of lacewings, ladybugs, and hover flies, among other hungry insects—as well as the occasional bird. The ladybug's orange eggs, nearly the same color and size as the resident aphids, were the hardest to spot. My well-established milkweed tolerated the aphid invasion fairly well, and I now grow it in infrequently used corners of my garden, protecting new plants until they can withstand the aphids feeding on them.

Certainly, I have killed my share of pests, insect and otherwise. I did not learn how to develop a beautiful, ally-friendly habitat overnight. Sometimes, I still step in to "disinvite" unwelcome guests. Fortunately, most pesky guests have natural predators. And a healthy garden is a haven for diverse life—plants, birds and other animals, fungi, microorganisms, and insects of all kinds. Such a garden is ultimately less work and better for the environment, and requires less, or ideally no, pesticides. As a bonus, by eliminating pesticides, we also protect the pollinators upon which we reply for much of our food.

Long ago, when I first began writing about natural enemies of herbivorous insects, my husband said to me, "Why don't you call them garden allies instead?" Perfect! And that's how I have come to think not only of the natural enemies of pests, but also the pollinators, decomposers, and other organisms that are part of any thriving garden. Now, I invite you to join me as I meander down the garden path, and meet some friends, both new and old—the garden allies.

INTRODUCTION

When we try to pick out anything by itself, we find it hitched to everything else in the Universe.—*John Muir*

Above, under, around, within, and beyond your plants, your garden is a constantly changing complex of connections. The garden food web is much like a spider web: gently pluck one strand and the vibration reverberates throughout the entire structure. Whether these organisms are friends or foes depends largely on how you, the gardener, perceive their place in your garden.

In these pages, we will explore a variety of common garden characters, large and small. We'll look at what they do and why, and examine their interrelatedness with one another and with garden soil and plants. You will emerge a wiser, more effective steward of your land.

It's not enough to just grow pretty flowers for our garden allies. Many natural enemies and other garden allies don't even need pollen and nectar, but have different habitat requirements that we can supply. Garden design is just as important as careful plant selection, as it incorporates all aspects of supporting the organisms in a healthy garden. Observing and understanding what you see is a start. Let's dig in.

ACTORS ON YOUR GARDEN'S STAGE

We will meet many garden allies in these pages, beginning with soil-dwellers and ending with birds and bats. You may learn things about them you never knew, including the myriad ways they can contribute to growing a healthy garden. Insects, however, are the stars of this book. Over 70 percent of this planet's animal species are insects (some estimates are even higher), and they are essential to its continued health.

But we cannot talk about insects without first talking about plants, the original solar collectors. They are the only organisms, with very few exceptions, able to store the sun's energy and convert it to food. Insects do a fantastic job of converting plant matter into food for other organisms. Thus, every animal on earth ultimately relies on plants for survival—sometimes directly (herbivores), sometimes indirectly (carnivores), and sometimes a bit of both (omnivores). Decomposers feed on all of the aforementioned. Additionally, most if not all insects are associated with at least one parasite that feeds on them; most have several. A majority of the following essays feature insects as leading actors in our gardens, as pollinators, decomposers, natural enemies, food for other insects and other creatures, and more. Everything, it turns out, is food for something else.

NATIVE VS. INTRODUCED?

Ecological concepts are woven throughout this book, for our gardens are truly ecosystems. As your understanding grows, you will find ways to take advantage of what the allies have to offer and to create or maintain conditions that welcome them. A key consideration in conservation biological control is that of "native" versus "introduced" plants.

What is biological control?

There are three kinds of biological control. Conservation biological control is the best method of pest management available to gardeners, and the focus of this book. By enhancing habitat (planting a beautiful garden) and simultaneously reducing or eliminating pesticides (healthier, less work, more fun), a positive feedback loop develops—and a garden will thrive with few interventions. Classical biological control is the importation and establishment of the natural enemy of an introduced pest (they are both introduced from the same part of the world). In augmentative biological control, a native or non-native insect is released as a short-lived solution.

Over the years, I have increased the number of locally native plants in my own garden. Why? Well, I value insects, birds, and other life they attract to my yard. It turns out that over 90 percent of herbivorous insects specialize on only certain plants. And they are most closely associated with plants native to their own environment, those with which they evolved. All those herbivorous insects attract and support predators such as ladybugs, lacewings, and many species of birds. Learn more about why we need plant-munching insects in our gardens (and thus native plants) in Doug Tallamy's eye-opening book *Bringing Nature Home*.

Take the oak tree, one of my favorites. Oaks are known to harbor over 800 insect species, including many species of caterpillars, and yet we rarely notice them. What do we notice? Birds! Almost all terrestrial birds feed their babies insects. And over half of those insects are caterpillars, most of which are more or less host-specific, feeding on a limited range of plants. Oak trees, where native, are essential sustenance for many species of the caterpillars the birds eat. Non-native plants, in contrast, often attract or sustain less local fauna, sometimes dramatically less. If you love birds—and who doesn't?—include plenty of native plants in your garden.

While I advocate for garden-friendly native plants, and there are excellent reasons to grow them, I am also pragmatic about gardeners. The truth is that few people plant entirely native gardens. We like to raise vegetables and to grow fruit, herbs, and beautiful flowers. If non-native hydrangeas remind you of your grandmother (they remind me of mine!), by all means plant some, even though they have negligible habitat value. Just be sure to include plenty of other plants that offer food and shelter to insects. A reasonable goal is to aim for at least half native plants. Many are attractive, and perform very well in home gardens. After all, they are well adapted to your area's climate and soil. Always select plants with care, especially non-natives—plants that are an asset in one place may be invasive in another.

The categories of "native" and "introduced" also apply to animals. In the best situations, native creatures of all kinds live in relative harmony with the native plants. But certain non-native animals and insects are also here to stay. They are not necessarily a serious problem, and humans (in agriculture as well as in home gardens) adapt. Some, such as the tiny parasitic wasps that were introduced to control cabbage aphids, are welcome garden guests. But just as non-native plants may come in, gain a foothold, and rampage unchecked through landscapes (for example, Japanese knotweed or common knapweed), so, too, can some non-native animals or insects (the brown marmorated stink bug comes to mind).

Non-native insect pests usually have few natural enemies present, and even if we have an established habitat garden, we may want to act. When we do, we should always choose the method with the least environmental impact. Poisons, organic or not, can have unintended repercussions. Do not react hastily. Figuring out an appropriate, and effective, response can require a bit of research in the books and websites recommended in this book's Recommended Resources. Try to understand what you are dealing with and what the best course of action might be, or, indeed, whether you must intervene at all. A resilient, self-regulating garden is a worthy goal!

ALLIES OR PESTS? GAIN PERSPECTIVE

The question often comes up: How can I tell a good bug from a bad bug? Of course, we know the pollinators are our allies, but what about all those other insects? I have a few tips, but first, I generally don't think of bugs as good or bad. Instead I have learned to think of them in their ecological roles: as prey and predators, pollinators, decomposers, and so on. Why? Well, "good" or "bad" depends on so much. For instance, some insects may be destructive in one region, and not in another, or present a problem in only one short part of their life cycle, while proving valuable in another. The caterpillar eating the leaves of my monkey flowers soon turns into a beautiful buckeye butterfly. The fact is, we need the herbivorous insects, and the entire complement of the food web, for a healthy, pesticide-free garden. Most of our significant "pests" were imported here (often unintentionally), where they don't have many natural enemies. Other insects have become pests due to drought or changes in climate.

I still use the term "pest" because sometimes insects are simply being destructive. One tip I can share is that if you only have a small population of a pest, it is unlikely to cause a problem. If you have an abundance of a particular species, learn its identity and life cycle before acting. Is it doing damage? What part of the plant is it damaging? How long will it be on your plant? Is it dining on a young plant, or threatening a prized centerpiece of your flowerbed? In a garden that has been designed to foster biodiversity, often we don't need to do anything. A predator shows up, and the problem abates.

Herbivorous insects

Did you know that insects are estimated to make up over 80 percent of all documented species on Earth, and that herbivores comprise between 25 and 35 percent of all insects? Without insects, the environment would collapse. Herbivorous insects have an essential role. They are crucial in the initial steps of breaking down plant material. And in the process, they become food themselves, for birds, amphibians, fish, reptiles, and other organisms that can't break down plants into digestible food. You can host plant-eating insects and still have a beautiful garden. It only requires a food-web perspective, coupled with thoughtful design and plant placement.

COMING TO TERMS

What is functional biodiversity?

Most gardeners think of biodiversity in terms of how many plant species live in their gardens. But all species—animals, fungi, bacteria, and more—contribute to biodiversity. How many individuals are there of each species? What is each species doing? Is it a pollinator, predator, prey, or decomposer? Are there other species that can take its place if it is impacted by a change or some unfortunate mishap? Functional biodiversity enhances the ability of a garden to be resilient and to self-regulate, bouncing back from disturbances. See Chapter 5 for a more complete explanation.

The truth is that less than one percent of insect species is harmful to our endeavors. Another small percentage benefits our crops and other human pursuits. Most are simply innocuous, going about their buggy business, oftentimes entirely unnoticed. "Pest" is also contextual. While I don't want aphids on my rosebuds, they are essential food for many animals, and there are places in my garden where I don't mind their seasonal presence. Without aphids, I ask children, what would happen to ladybugs? A world without them is unimaginable, of course, and I once caused a kindergartener to burst into tears when he realized the dire consequences of my question.

Keeping the food web in mind provides important context. For example, I once watched a tachinid fly larva emerge from a monarch caterpillar. It was gross, and fascinating, and I was a bit sad for the butterfly-not-to-be. It made me reflect, as I often have, on our attitude toward insects. The longer I study nature, the harder it is to find an easy answer to "how can I recognize the good bugs?" (Though I admit that I once taught a class titled "The Good, the Bad, and the Bugly"—I couldn't resist.) No doubt, sometimes insects behave as pests, when they are damaging plants beyond an acceptable threshold. But an overall healthy ecology includes both prey—often herbivorous insects—and predators, and a shift in how we gardeners perceive them.

THE TROUBLE WITH PESTICIDES

Gardeners who want to reduce, or better yet, eliminate pesticide use (all of us, right?) would do well to learn about "conservation biological control," the technical term for the art and science of enhancing natural-enemy populations, which I will explain in more detail in the pages ahead. Use pesticides minimally, if at all, and only as absolutely needed, first trying more benign methods. For instance, you can wash off an aphid infestation with a blast of the hose, and protect susceptible vegetables from marauding caterpillars or leaf miners with row covers.

Patience is a virtue with conservation biological control. Well-established gardens self-regulate more reliably when it comes to their inhabitants. If your garden is brand new, or you are transforming a former lawn to suitable habitat, it may take a while to reach that stage. In the meantime, use only mild organic pesticides (I am a fan of Sluggo and Safer insecticidal soap products), and carefully administer only to targeted areas.

The saying that an "ounce of prevention is worth a pound of cure" applies here. Your garden is less likely to develop problems if you have a diversity of plants, grow them in good, healthy soil, and pay attention to their needs and condition. Also, you will have better luck when you select plants that are regionally adapted and robust, and eliminate those that are prone to infestation. Non-native plants are often more prone to pests and disease. Gladioli, hibiscuses, and crape myrtles are good examples (some may be billed as pest- and disease-resistant, but many are notoriously vulnerable). Do some homework before purchasing plants, not only on the internet and in books, but also from local resources, such as garden clubs or the staff of a trusted garden center.

The truth is, pesticide use often backfires. This is because some individuals of the pest survive to reproduce. After a few generations, they become resistant to the pesticide. Secondary outbreaks, sometimes worse than the initial one, can occur. By now, predators—and I don't just mean other bugs but perhaps reptiles, birds, fish, and other animals—are also sick or dead, having ingested the poison from their contaminated prey. Also, pesticides can kill the vulnerable pollinators we treasure. In short, pesticide use in your garden can wreak unintended ecosystem havoc.

As you use less pesticide and improve habitat, natural enemy numbers will increase and you will discover that pest species become less problematic. It's a positive feedback loop. After a while, you may wonder why you ever reached for those "solutions."

CAN PURCHASED BENEFICIALS REALLY HELP?

There are certainly times when purchasing insects is the best approach to a problem. A release of parasitoid wasps may be exactly what is needed to save a prized, mature tree from an invasive insect. Be aware, however, that many purchased biological controls will not be effective in a home garden. Some only work on a large scale. An example is the protozoan *Nosema locustae*, which farmers use for grasshopper control. Some simply won't work as intended—they are mainly sold because of consumer demand. I'm sorry to say that ladybugs are an example (more on that in Chapter 4).

Purchased lacewings, on the other hand, establish resident populations when introduced to gardens. They dine on unwanted soft-bodied insects and more (full details are in the essay devoted to them). There are biocontrols that can be purchased for specific needs: beneficial nematodes (which attack soil-dwelling larvae of unwanted insects and pest nematodes), and bacteria such as *Bacillus thuringiensis* (which kills caterpillars; target carefully so desirable species are not harmed). But the truth is that if you have fostered plant biodiversity and provided for other needs such as water and shelter, you will rarely need to purchase insects or import remedies.

DO WHAT YOU CAN DO

Insects are key partners in maintaining the web of life and in supporting healthy gardens. Without insects, and the intricate relationships that connect them to plants and to other organisms, all life would ultimately collapse. By some estimates, a third of every bite of food we eat is the result of pollination—almost exclusively by bees—but the insects that are food for other animals, and those that regulate pest populations, also play a vital role.

Climate change, pandemics, and other stresses of contemporary life have increasingly turned us to our gardens for solace, for recreation, to create wildlife habitat, and to provide food for our tables. Now is the time to learn how to garden in a manner that supports continued life on the planet. The alarmingly sharp decline in insect populations, the "insect apocalypse" in the news, is not some abstract notion, but a stark reality if we do not act.

In the end, it is up to us, as *human* garden allies, to commit to environmentally sound practices (and to encourage our gardening friends and neighbors to adopt them). Nurture your soil, the foundation of any garden. Learn the design principles of ecologically healthy gardens. Install a bat box to help control mosquitoes. Bundle up some hollow reeds and tuck them in a protected place for native bees and solitary wasps to nest. Include some thorny shrubs for nesting birds, or a home for toads. Stop using pesticides and other garden chemicals. Choose native plants, not only milkweed for monarch butterflies, but plants for the other insects and native animals that rely on them. Create wildlife corridors with "habitat hedgerows" instead of traditional shrub borders. Grow some of your own food, start a worm-composting bin, and learn about seed saving.

Beyond your own property, you can do even more. Urge your local nurseries to stop selling regionally invasive plants. Participate in your neighborhood community garden. Join a local native plant society and botanic garden (and make some new friends). Volunteer there, or at your local school garden—they teach the next generation, and they need your help!

USING THIS BOOK

As I have mentioned, the majority of our garden allies are insects, so much of this book is about them, but we will explore many other garden organisms. While each essay can be read as a stand-alone, I encourage you to begin at the beginning, as new terms are introduced as the book progresses. I always define new terms; you can consult the Glossary and Index in the back of the book as well. Also, as you'll see, the essays flow from one to the next. Some feature one small ally group, while others are much broader in treatment. Each finishes with a brief summary of pertinent facts and taxonomic information.

While telling many intriguing stories, and unveiling some occasionally surprising relationships (really, we need aphids!), this book alone is not enough. Do you have more questions about insects as you read this book? I hope so! I keep BugGuide.net handy for all my insect questions (it covers other terrestrial arthropods as well). Books, websites, podcasts, and online-streamed programs and training that promote ecologically sound gardening practices can also guide you and expand your knowledge. Good regional insect field guides are helpful, especially if they include photographs of different life stages. A small ruler may come in handy (and metric conversions are provided in the back of the book) and a good hand lens is indispensable for anyone interested in their garden's microcosm.

Information and advice herein has been much improved by the dozens of scientists who have made suggestions and corrections. Their expertise is reflected throughout this book. It would take an encyclopedia to share all the amazing garden allies and their fascinating stories and interactions. I also maintain a Facebook page (facebook.com/Garden.Allies), where we can share tips and photos. This book, I hope, is but a beginning.

A
CLOSER LOOK
AT "BUGS"

We fall into the habit of using the term "bug" to include all the small creatures we meet, but this really only properly refers to one order of insects. Admittedly, though, "bug" is convenient and widely understood, and I occasionally use it instead of "insect," especially when I am referring to other arthropods and invertebrates along with insects.

The "true bugs" are classified in an overarching order called Hemiptera, distinguished by having piercing, sucking mouthparts (meet them in Chapter 5). True bugs are nested within the broader category of insects. Insects have three main body parts (head, thorax, and abdomen), three pairs of legs, one pair of antennae and, usually, two pairs of wings. Insects are themselves nested within the arthropods, which includes crustaceans, centipedes, millipedes, spiders, scorpions, mites, and ticks, and a few other interesting organisms—many are described in this book. Arthropods are nested within the invertebrates, animals without backbones, which include some other creatures we will meet ahead, such as earthworms, snails, and slugs.

COMING TO TERMS

What is an arthropod?

This is a word you will see often in these pages, and wherever gardens and their inhabitants are being discussed. Arthropods, animals with an exoskeleton, have segmented bodies and jointed appendages. Their exoskeleton must be shed ("molted") in order to continue growing. In addition to the world's many insects, other major groups of arthropods include crustaceans, which are mostly marine, myriapods (centipedes, millipedes, and relatives), and the chelicerates. Arachnids, which include spiders, mites, and ticks, are all chelicerate groups, as are scorpions and a few other forms not likely to be found in gardens.

CHANGING APPEARANCES AND LIFE STAGES

One thing to keep in mind when learning about insects is that they may have different appearances, behavior, and diets in different phases in their life cycles. The majority undergo "complete" metamorphosis: the familiar sequence of egg, larva, pupa, and adult ("*holo*metabolous"). Others undergo "incomplete" metamorphosis: egg, nymph (or naiad for aquatic forms such as dragonflies), and adult, in which juvenile forms resemble adults ("*hemi*metabolous"). While hemimetabolous insects develop wings externally—think of a juvenile grasshopper's short wings—holometabolous insect wings develop hidden from our sight, in the pupal stage.

We often use unique names for the different phases of insect life cycles to reflect their changing appearance. Thus caterpillars are the larvae of moths and butterflies, grubs are beetle larvae, maggots are fly larvae, and nymphs and naiads are commonly used terms when one is speaking of the "hemimetabolous" insects. Gardeners are sometimes amazed when they connect the dots and learn that this caterpillar is that butterfly's earlier stage.

Mouthparts, an important feature used to classify insects, are the same throughout the life of hemimetabolous insects. With some exceptions, they generally eat the same food from juvenile to adult stages. Good examples of this are grasshoppers and true bugs. Holometabolous juveniles and adults, on the other hand, may feed on completely different food sources, and have mouthparts best adapted to each life stage. Thus caterpillars have chewing mouthparts and eat leaves, while the adult butterflies they become have siphoning mouthparts and are nectar feeders. How insects feed, of course, affects how we perceive their place in our garden, and how we may manage their presence.

HOW IS IT NAMED?

Scientific nomenclature can also give insight into types of organisms. For instance, knowing that in Greek "hemi" means "half" and "ptera" means "wing" sheds light on the meaning of Hemiptera and the two historical suborders, Heteroptera (different-winged) and the former Homoptera (same-winged).

I've provided common names if the animal described has one—many do not, especially among the insects. Some are descriptive, as in "paper wasp" and "roly-poly." You might be intrigued to learn that there is a logic to many of these common names. When a bug's common name is written as two words— for example, assassin bug, stink bug, and ambush bug—it belongs to the above-referenced Heteroptera, the "different-winged" insects. When "bug" appears as the end of a single word, however, the animal generally (but not always) belongs to a different group. For instance, ladybugs are beetles (Coleoptera), doodlebugs are antlion larvae (Neuroptera), and pillbugs aren't even insects, they are crustaceans. Incidentally, the same rule holds true for other insect orders. Hover (syrphid or flower) flies and house flies are true flies, Diptera, while a dragonfly is in the Odonata, and butterflies are in the Lepidoptera.

CLASSIFICATION BASICS

Let's pause for a moment and orient ourselves to scientific nomenclature—the "formal names." The nested categories scientists use have conventions, and word endings indicate the category levels. Insect orders most often end in –ptera, meaning winged, for example, Lepidoptera, which means scale-winged. Here's a quick example of how scientific classification of the spicebush swallowtail butterfly breaks down:

Kingdom: Animalia

Phylum: Arthropoda

Class: Insecta

Order: Lepidoptera

Family: Papilionidae

Genus: *Papilio*

Species: *Papilio troilus* (this, the scientific name, is called a "binomial" because it's traditionally composed of two words, the genus and the specific epithet)

Common name: spicebush swallowtail

The binomial identifies a creature and always includes the genus (plural genera) followed by the specific epithet. Together, these describe the species. If the species is not certain or is unknown, or sometimes for the sake of brevity, it may abbreviated as sp. (singular) and spp. (plural). Returning to the example above, we can say quite precisely that the butterfly is *Papilio troilus*, or, less precisely, refer to it as *Papilio* sp.

Super- and sub-orders are not uncommon, and can be spotted by their word endings. When I discuss insect families, I often use the less formal family name, with an –id ending, and not capitalized. Thus, in referring to our Papilionidae butterfly example, it would be accurate to call it a "papilionid." Consult the box at the end of each essay for fuller taxonomic information.

In talking about animal taxonomy (or plants for that matter), the term "taxon" (plural, "taxa") is often used. It applies to any rank. For example, I might say "flies, beetles, moths, and butterflies, and the wasps, bees, and ants are each in their own taxon" or "common worm taxa include. . . ."

I use common names in the following pages, when they exist, and supply scientific names. (I have not always included the scientific names of plants, except where needed for clarity.)

Note: As we learn more about genetic relationships through DNA testing, taxonomists rearrange organisms into different related categories. Some things we can count on—one of those is name changes. Scientific names used here were current at time of publication.

FORM OR FUNCTION?

For our purposes as curious gardeners, it is rarely necessary to identify insects down to the species level. It can be a great challenge even for the experts. Most entomologists specialize in just one insect. And I don't mean beetles, or even one family of beetles. Many are specialists in a single genus. This level of specialization still gives scientists plenty to study! An entertaining, insider picture of that realm and its ongoing progress can be found in paleontologist Richard Fortey's wonderfully readable book *Dry Storeroom No. 1: The Secret Life of the Natural History Museum.*

As I consider the animals sustained in my garden, I have found it helpful to think not just in terms of what a given creature *is*, but what it *does*—this is the biological concept of "guild" (a group of organisms exploiting the same resources). Among our garden allies (and beyond), there is a flower visitor guild, a pollinator guild, a predator guild, a decomposer guild, etc.

You will meet many flower visitor guild members in these pages. There is often overlap—that is, many insects and other organisms appear in more than one guild. Not all flower visitors, for instance, are effective pollinators. The pollinator guild includes numerous insects, especially bees, flies, and butterflies. But the pollinator guild also includes hummingbirds and bats—animals that are not even closely related. Hover flies belong to flower visitor and pollinator guilds, but their larvae are in the predator guild. I like this way of thinking, because scientific names can change as we learn more about insects, but guilds that reflect ecological roles remain stable.

Now it's time to get into the garden and see what creatures we may discover there. I invite you to go out and watch the insects visiting your flowers, gently lift a rock to see what is living underneath, get out the binoculars and observe the birds splashing in a birdbath. Explore your garden, and you will become more familiar with the garden allies you will meet in these pages.

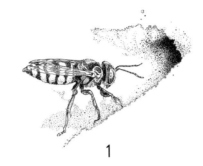

1

LIFE BENEATH OUR FEET

The air into which plants extend their stems, leaves, flowers, and fruit is a virtual desert compared to the soil in which their roots seek anchorage, water, and nourishment. This is why the health and sustainability of any landscape begins with stewardship of its most valuable resource—the soil underfoot.

Soil is not simply a canvas on which we paint our beautiful plant picture, but a living substrate. A good gardener, before anything else, tends the soil, the foundation of the landscape. We think that we are growing plants, but really we are growing soil. When we proudly show off our prize tomatoes, we could just as proudly show off a handful of the fertile soil from whence they came. Fertile soil includes a complement of humus, organic matter that has decomposed until it has reached a stable state. We have many garden allies, seen and unseen, to thank for that.

As Leonardo da Vinci remarked long ago, "we know more about the movement of celestial bodies than about the soil underfoot." It is still true today. Just about every available surface under our feet is teeming with life. Incredibly, even the thin film of water that coats soil particles and lines pores in the soil harbors microscopic organisms that navigate those minute spaces.

Fascinating stories emerge from the soil—stories about creatures as small and numerous as bacteria, as large as giant earthworms, and as musical as crickets. A mere teaspoon of soil may harbor over 1 billion bacteria. Most are beneficial; some bacteria can even break down pesticides and other pollutants. Oregon and Washington are said to be home to giant (but elusive) earthworms that smell strongly of lilies. To attract mates, Jerusalem crickets play the "drum" by beating their abdomen on the ground; each species produces a characteristic song.

We begin our exploration of garden allies with the mysterious goings-on beneath our feet. From microscopic, often bizarre-looking creatures to more familiar animals such as earthworms, the extraordinary soil food web is the basis of our gardens.

Earthworms: Wigglers to Nightcrawlers

Every garden includes a free fertilizer factory. A single earthworm may produce up to 10 pounds of nutrient-rich castings (worm manure) per year, while also improving soil structure and drainage.

D id you know that Charles Darwin was an earthworm expert and is largely responsible for bringing them to our attention and appreciation? Of course, he is best known for his theory of evolution by natural selection, but he studied many natural history subjects in depth. Orchids, carnivorous plants, barnacles, and coral reefs all fell under his encompassing scrutiny. In the case of earthworms, he devoted over 40 years to studying them, from shortly after he quit sailing the Beagle until near the end of his long life. During the time when Darwin was lying on his belly watching towers of castings forming on the surface of a pasture, worms were generally regarded as pests. And until the publication of his seminal book *The Formation of Vegetable Mould, Through the Action of Worms* in 1881, earthworms were not even deemed worthy of scientific attention. Now, thanks to him, we understand they are an important part of the soil food web. Indeed, Darwin is regarded by many biologists as the founder of soil science.

Gardeners now hold earthworms in high regard. But even so, few of us understand the full scope of their role, not only in the garden but in field

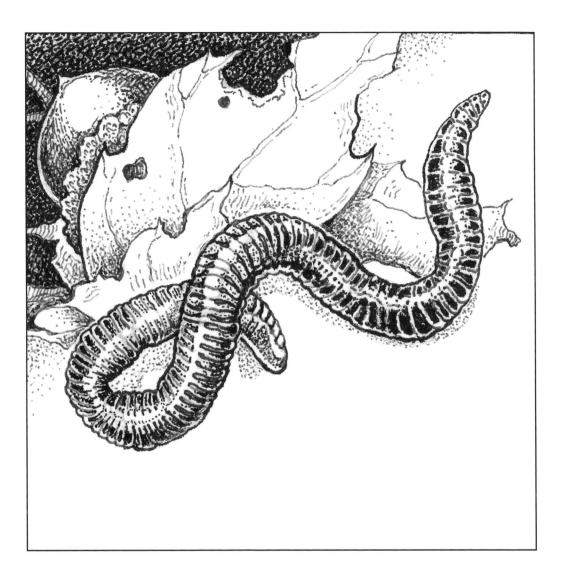

Red wiggler *(Eisenia fetida)*

and forest as well. You may know that they literally eat soil. Did you know most species also include plant material in their diet? They prefer to feed on that which is already decomposing (a process started for them by microbes). Also, many earthworms can eat their own weight in organic matter daily—can you imagine? By comparison, the average human consumes only about 3 to 5 pounds of food each day.

Earthworms benefit soil in other ways, too. Their earthwork is often impressive as they accelerate decomposition in the soil. Their tunnels create channels for penetration of roots, water, air, and nutrients. They can be especially useful in aerating compacted soil and save you the work of doing it yourself with equipment or back-breaking digging. Furthermore, their castings include soil particles and digested organic matter coated with mucus. This in turn creates aggregates that concentrate humus and improves soil structure. This is a rich source of nutrients in your garden ground.

Considering how important they are, let's talk about garden practices that could harm these beneficial creatures. Tilling the soil is usually detrimental, as it can kill worms by cutting or chopping them up as well as collapsing their tunnel networks. It also destroys the soil layers, each of which harbors its own community of organisms. To increase earthworm populations, and develop healthy soil, I recommend no-till techniques. Build new garden beds by "sheet mulching" (cover compacted soil with layers of compost, mulch, and/or a cover crop). Wait a few months while the earthworms do their work, then plant in the enriched and loosened soil. (For more information on no-till methods, along with some important exceptions, I recommend Lee Reich's outstanding book *Weedless Gardening*). Do not use garden chemicals, which can harm earthworms and other soil residents.

You can also invite or encourage earthworms in established garden beds simply by adding compost. For plants that need rich soil, I deposit a thin layer of compost as mulch twice a year and, as mentioned above, let the worms do

Nightcrawlers (*Lumbricus terrestris*) mating

the work. If your soil is very compacted, and you don't have time to wait, go ahead and till as a one-time thing. Studies suggest that soil enriched with worm castings produces plants that are better able to resist pests and diseases—yet another good reason to welcome earthworms.

Earthworms are divided into three main groups. *Epigeic* worms, which include the common red wiggler used in compost bins, live on the surface of the earth and eat rotting organic matter. *Endogeic* worms live within the soil, build lateral burrows, and are the only worms that eat large quantities of soil. They have little effect on surface litter, preferring soils rich in organic matter. *Anecic* worms, also known as nightcrawlers, come to the surface at night to drag leaves and other organic matter down into their deep and permanent vertical burrows. There, the scavenged materials break down and release nutrients. Thus anecic worms play a central role in the decomposition of leaf litter and soil formation. It is safe to say that all earthworms have a profound impact on the soil food web wherever they are found.

Interestingly, these creatures have both male and female reproductive organs (that is, they're hermaphrodites). Some are able to reproduce asexually ("parthenogenically"). Earthworms produce one to several eggs enclosed in a cocoon. Red wigglers may produce over 900 eggs annually, while nightcrawlers produce a mere 40 or so. Upon hatching, young worms take six months to two years to mature. Red wigglers live about a year; nightcrawlers may live as long as six years.

While earthworms contribute fertility to our gardens and farms, most, including nightcrawlers, are non-native species in North America. Although native earthworms are found in some regions of the United States, a large area of North America had no earthworms at all since the last period of glaciation over 10,000 years ago (and perhaps not before), until their introduction by early European colonists.

The forest understory in parts of Minnesota is disappearing due to non-native earthworms' voracious appetite for leaf litter. These areas naturally had loose soil, with a thick layer of duff (decomposed organic material). Researchers found that the worms' actions actually compacted the soil and created an environment unsuitable for native forest forbs. Even in regions with native worms, non-native earthworms often dominate, displacing native species.

In recent years, mainly in the Midwest, "jumping worms" have invaded. They are dark gray and otherwise look much like nightcrawlers, but they sport a telltale white band around their middle. Jumping worms belong to several species in the Asian genus *Amynthas*. They earn their name. If you come upon or handle one, it will wiggle in a wild snakelike motion and may even leap into the air or detach the tail-end of its body in a frantic effort to escape.

These deprive forest trees, especially young ones, of essential nutrients. They aren't likely to be good news in the garden, either. Keep a lookout and report sightings to the nearest Cooperative Extension office. In particular, watch for hitchhikers in soil deliveries, leaf piles, and newly purchased potted plants.

Once an earthworm species becomes established, there is little one can do to control it. While not a problem for your garden or a farmer's field, where earthworms are already established, it could become so in wild settings. Therefore when fishing, never dump leftover live bait, which is mostly non-native worms.

Finally, in addition to the role of most earthworms in improving soil, they also serve as a valuable food source for birds, moles, and other organisms. A garden without earthworms is simply unimaginable!

APPEARANCE Typically red wigglers are 2–5 in. (50–130 mm) and reddish brown; red leaf worms are 2.5–6 in. (60–150 mm); nightcrawlers are 4–12 in. (100–300 mm) and pink to dark red. With few exceptions, immature earthworms look much like adults.

POPULAR NAMES Earthworm, nightcrawler, jumping worm.

SCIENTIFIC NAMES Phylum Annelida (segmented worms, leeches, polychaetes), class Oligochaeta, family Lumbricidae, which includes most familiar garden earthworms.

COMMON TAXA Red wiggler (*Eisenia fetida*), red leaf worm (*Lumbricus rubellus*), nightcrawler (*Lumbricus terrestris*).

DISTRIBUTION About 180 species in 12 families in North America; over 30 percent introduced. Worldwide, about 4000 different species.

GARDEN ACTIVITIES Decomposers; they also aerate and improve soil.

Microscopic Decomposers: Slime Molds, Protists & Others

Breaking down organic material—and rocks!—is the impressive work of a miniature army of microscopic organisms. The result? Soil!

Without living organisms, soil would be merely crushed rock. Interactions between microorganisms and mineral components, fueled by sunlight and chemical reactions, work to break down rock. Over time, soil develops. This process is slow but remarkable.

Bacteria play an essential role in creating and building up soil. Single-celled and not visible to the naked eye, these minute life forms break down organic matter, converting inorganic matter into organic forms, and serve as food for other organisms. It was 19th-century biologist Louis Pasteur who first demonstrated that the initial decomposition of organic material was achieved by bacteria, and not (as was supposed) as the result of a chemical reaction. All the kingdoms of life are involved in the decomposition of organic matter, making nutrients available to plants and ultimately animals. But only bacteria are able to make all the essential elements available. And bacteria are numerous, to say the least. As mentioned earlier, there can be over 1 billion in a teaspoon of soil.

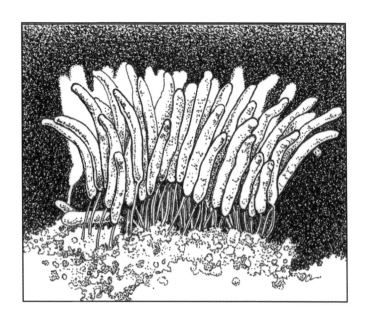

Most plants, algae, and some bacteria and protozoans provide their own nourishment (termed autotrophs), producing complex organic molecules using light energy or chemical reactions. They literally create the basis for life, generating the organic molecules necessary to support more complex life forms. Their role is absolutely critical. All organisms that rely on organic matter for nourishment are heterotrophs; this includes the myriad of organisms that contribute to soil formation.

Here's how it all unfolds. Begin with bare rock— let's imagine the volcanic Hawaiian Islands when they first emerged from the sea, for instance. The first organisms to colonize land newly created by lava flows had to be able to provide their own nutrients by means of light or chemical energy (autotrophs), as there was no other food source. The very first colonizers were blue-green algae (now known as cyanobacteria). They are able to photosynthesize, and some are also able to "fix" atmospheric nitrogen and make it available to plants.

When certain algae or cyanobacteria combine with specific fungi, lichens, which are powerful rock decomposers, are formed. They gather moisture and nutrients from the air and, over the years, bits of the rock break off and sift or fall to the ground (or are moved there by wind and rain), contributing a small but steady delivery of minerals and nutrients. In colder climates, this process is accelerated by the freezing and thawing of water over successive winters. Lichens also produce unusual acids that help break down rock.

Through the action of such agents, eventually a thin layer of soil develops on an initially barren surface like the Hawaiian volcanic rock. Next, higher plants move in; many of these have a nitrogen-fixing capability (I will explain more about this in the following essay). As more complex flora develops, more actors may get involved and what we call "the soil food web" becomes established. These include decomposers, scavengers, and recyclers, such as slime molds, protozoans, lichens, a variety of fungi, and other more obscure organisms.

When soil food web organisms are abundant and able to break down organic matter, soil builds up. Water-holding capacity improves, nutrients increase. As organic matter breaks down, that "black gold" called humus, so desired by us gardeners, is created. Humus is a stable end product, organic matter that is resistant to continued decomposition. It contributes a crumbly, spongy texture and a rich, dark color to soil. Voilà—with time, the process kick-started by tiny organisms results in a fertile, soil-covered planet.

That's not the whole story. These denizens of the soil food web are part of a complex system. A wide variety of predators, parasites, and pathogens are also present and regulate populations. In addition to predator, parasite, and herbivore, other examples of soil food web roles include organisms that feed on fungi (fungivores), on dung (coprophages), on plant detritus (detritivores),

Testate (shell bearing) amoeba (*Euglypha* sp.)

and on dead plants and animals (saprophages), as well as bacterial and fungal partners that assist in nutrient and water uptake by plant roots. Lifespans of these organisms vary quite a lot. Soil bacteria and protozoa are short-lived, but lichens and some fungi may survive for many years.

Variety abounds. *Archaea* (until recently classified as bacteria) were once thought to occur only in extreme environments such as hot springs and saline soils, but have now been found in association with roots in more ordinary environments. Some bacteria are found principally around plant roots, where they find a nutrient source in root exudates and dead cells. Actinomycetes, a bacteria, are a source of antibiotics such as streptomycin, tetracycline, and actinomycin, and are a principal contributor to the earthy aroma of forest humus and well-made compost.

Categorizing all these many and minute creatures is a challenge and a subject of ongoing research. Yellow, brown, green, blue-green and red algae belong in different groups. Fungi include yeasts and mushrooms. Water molds are in their own group, the oomycetes. Other odd organisms, such as chytrids

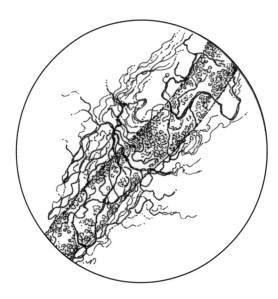

Actinomycete bacteria on plant roots

and hypochytrids, are also members of the soil food web. For convenience, "umbrella" groups have been named to help us talk about the strange, the outliers, and the things that do not fall easily into categories of plant, fungi, or animal. For instance, protists are unrelated single-celled or multicellular organisms with no specialization of tissues. Protozoans are animal-like protists such as amoebae.

Let's return to some that affect gardeners the most, the decomposers. The sugar fungi (saccharomyces) include the first fungi to attack dead leaves. Slime molds, sometimes similar in appearance but unrelated to fungi, help to break down leaves and wood by engulfing food in their path. In some species individual cells are able to join forces (to "aggregate"). Protozoans patrol the film of water in soil pores, also regulating the bacterial community. Unfortunately, some bacteria, fungi, and other organisms found in soil are agents of plant diseases. Oomycetes, for example, cause sudden oak death and diseases such as potato and strawberry blight.

How can you help nurture the good and desirable denizens of the soil food web in your home garden, especially when you can't even see them? Don't use pesticides, including herbicides, which harm them. Limit soil disturbance and tilling. Where feasible, allow leaves and other plant materials to decompose in the garden. If you are successful, your garden soil will be organically rich and support a variety of thriving plants.

As we turn our attention to the soils in which our gardens grow, we discover a new world to explore. Closely examined, the discovery leads to a new perception of the gardener's footprint in the garden, and a joy in the wonders of life that surround us every time we step out our door.

APPEARANCE Varied and diverse; the majority are visible only through a microscope.

POPULAR NAMES Bacteria, protozoans, algae, diatoms, slime molds, fungi, lichens, and water molds.

SCIENTIFIC NAMES The historical five kingdoms of life are obsolete. Taxonomy of these groups is complex, and under revision. See tolweb.org for further information.

COMMON TAXA Bacteria, protozoa, slime molds, fungi, actinomycetes, lichens, water molds, sugar fungi.

DISTRIBUTION Worldwide. Most species of soil microorganisms remain undescribed!

GARDEN ACTIVITIES Decomposers of organic (leaf litter, dead animals) and inorganic (rock) materials.

Mycorrhizae, Nitrogen-Fixers, Mushrooms & Other Nutrient Facilitators

Did you know that the mushrooms in your garden are akin to fruit on a tree? They are the fruiting body of the fungus, popping up from an extensive network of "roots" (hyphae). Some edible species can even be introduced into our gardens.

While many soil organisms are involved in soil-building through decomposition, members of another largely invisible group facilitate nutrient uptake by plants. In case you ever had any doubt that the natural world is interdependent, look here. It's all around us and it's fascinating.

In this mutualistic realm between soil life and plant life, the importance of activity involving nitrogen cannot be overstated. Why is nitrogen so key, you may ask? Well, we know that nitrogen gas makes up a whopping 79 percent of our atmosphere. It also turns out to be an essential component in the building blocks of life itself such as DNA, proteins, and chlorophyll molecules.

The problem is plants on their own cannot avail themselves of nitrogen. It must first be converted, or "fixed," into a form that can be taken up by roots. In nature, this critical process is primarily accomplished by soil microbes and

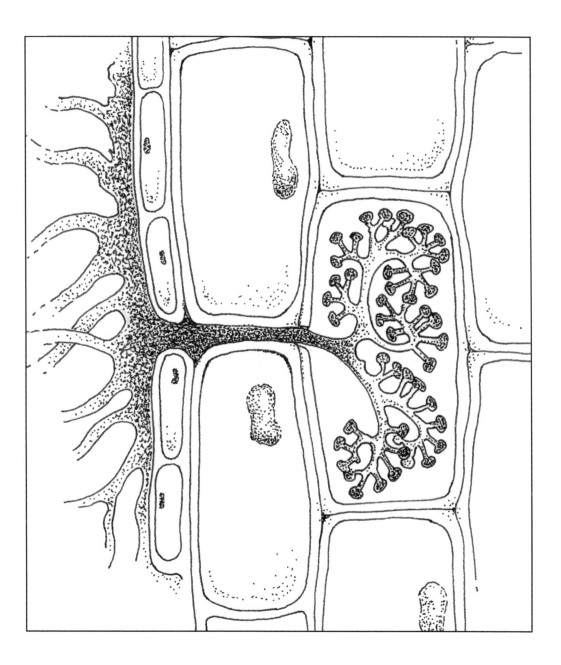

Endomycorrhiza

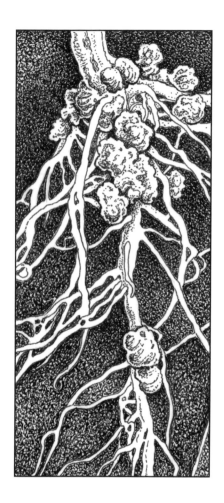
Root nodules on legume

fungi. In return for providing nutrients and water, fungal and bacterial partners are nourished by root exudates such as sugars—a fair exchange.

Among these actors are mycorrhizal fungi, rhizobium, actinomycetes bacteria, and some cyanobacteria. Certain fungi and bacteria form "obligate mutualistic" relationships with plants, a relationship in which both rely on the other for essential sustenance. They can't grow at all without their unseen partners! Other nitrogen-fixing bacteria, the cyanobacteria, live free in the soil or water. Some have plant associations, such as a cyanobacteria that fixes nitrogen in the water fern *Azolla*, contributing fertility in Asian rice paddies. Roaming or completely interdependent, these invisible partners provide plants with necessary nutrients in available form, and sometimes also increase water uptake.

To cite the most familiar garden example of these partnerships, let us turn to peas and beans (and other related leguminous plants). Pull up a bean plant and the roots, if well-inoculated with bacteria, will be covered with pale pinkish or white nodules, technically a type of gall

(Chapter 6 includes an essay devoted to galls). Specifically, you are looking at nitrogen-packed cells produced by *Rhizobium*. The nitrogen becomes available to a subsequent plant.

There is a slight catch. For the nitrogen contained in the nodules to become available, the roots must decompose. Over the cycle of a gardening year, this will happen naturally in our vegetable patch with our bean-family plants. It also occurs in nature in, for instance, Maine's famous roadside lupines or the state flowers of Texas, the bluebonnets. We can thank the unseen partners that these glorious leguminous wildflowers are able to return and often increase their numbers from one year to the next.

Nor are leguminous plants the only beneficiaries of such relationships. A taxon of actinomycetes bacteria, the *Frankia*, form large clusters of nitrogen-fixing nodules on the roots of bayberries, alders, mountain mahogany, and some ceanothus species. Botanists are looking into this further; so far they've identified 200 such plant species in 8 families. Accordingly, these are called "actinorhizal" plants.

Fungi, too, help many larger plants. But before we investigate, it is important to note that fungi are actually not plants. It turns out their cell walls are not made of cellulose, as in plants, but of chitin, as in animals. Also, plants generally make their own food, while fungi do not.

Fungi are commonly mycorrhizal. These have a mutualistic relationship with certain plants. The fungi use sugars produced by the plant's roots, while increasing the availability of water and several nutrients, most notably phosphorus, to the plants. Incredibly, they are reputed to increase roots' effective surface area by a factor of 1000! This relationship was first documented in 1885 by a German scientist, Albert Bernhard Frank, in a simple experiment comparing pines grown in sterile soil to pines grown in forest humus. You can guess which pines prospered and which did not. He identified this fungus as the beneficial agent for the healthier pine trees, specifically a type called "*ecto*mycorrhizae" that are associated with conifers and hardwoods. These inhabit a web surrounding the cells in the roots of their host.

The fruiting bodies of ectomycorrhizae are the familiar mushrooms of our woodland hikes—edible fungi such as *Boletus, Cantharellus,* and *Morchella,* and poisonous mushrooms such as many *Amanita* species. They are specialists on certain species of plants, and any good mushroom forager knows which tree species are associated with their favorite fungi.

Yet another variation are the fungi species termed "*endo*mycorrhizae." Unlike their relatives, their microscopic growing branches ("hyphae") are able to penetrate through the cell walls of their hosts. They are generalists and may be found on many unrelated plants. Perennials, softwood trees, shrubs, and many vegetables have been discovered to be associated with these facilitators. A notable exception is that they do not associate with members of the large mustard family, which includes garden favorites like broccoli and cauliflower as well as beloved flowers such as nasturtium. We don't know why—another instance of the mysteries of nature.

The more we study mycorrhizal relationships, the more we discover how ubiquitous they are. Some plants may host dozens of species and include both types. We now realize how essential soil fungi and bacteria are, though we have much more to learn. Foresters use the clever term "wood-wide web" to describe the complex relationships between the trees and the earth. The widely hailed 2019 documentary film *Fantastic Fungi* riveted audiences with its compelling images and information about what is going on under and in the ground between fungi and plants.

Now, how can we avail ourselves, or rather our plants, of these important and beneficial relationships? Well, unless the soil in your garden is impoverished, native endomycorrhizae should be present; purchasing any is rarely called for. *Rhizobium* and ectomycorrhizae, however, may need to be introduced. One example is a vegetable patch where you want to grow green beans

Honey mushroom (*Armillaria ostoyae*)

or peas. Such a site may benefit from the addition of legume-specific commercial inoculants to boost microbial activity (some other formulations are for more general use). These come as a granular substance, sold in bags at garden stores and online suppliers; follow the application and timing instructions with care for best results.

For many years, I have increased the overall fertility of my vegetable garden with a fall crop of fava beans. When the plants begin to flower in spring, I chop them down at ground level (sparing a few so I can harvest the delicious edible flowers and beans). I bury the tops in the compost pile and leave the fibrous root system to decompose in place. No tilling necessary! This project is always worthwhile in a vegetable patch, but also if your soil is poor and infertile, such as when you start a garden in a new home where construction activities removed or compacted the existing topsoil, or where you wish to convert an area of lawn or groundcover to a new vegetable garden.

Mycorrhizae can make a difference in a new garden or bed with depleted, poor soil, or even added to a container or window box of sterile potting mix. It has long been my practice, when planting native shrubs in my garden, to gather a handful or two of humus from the nearby woods and use it to inoculate the planting area. A walk in the woods is never in vain, and it is always rewarding to pull the aromatic bundle from my pocket and tuck it into my garden.

Lastly, a note on the lovely lupines. If you've had poor luck with them, here's a time when you may consider adding a legume-specific inoculant powder to mix and coat the seeds prior to planting—you guessed correctly, it contains the right kind of mycorrhizae.

APPEARANCE Nitrogen-fixing *Rhizobium* and *Frankia* are found in plant-formed nodules, often in clusters, on roots. The fruiting bodies of ectomycorrhizae appear aboveground as the familiar mushrooms we find in association with plants; endomycorrhizae are not visible to the naked eye.

POPULAR NAMES Mushrooms, nitrogen-fixers.

SCIENTIFIC NAMES Kingdoms Fungi and Bacteria. Bacteria discussed here include *Rhizobium* and *Frankia* species.

COMMON TAXA Mushrooms, ectomycorrhizal fungi that grow in association with plants, usually hardwoods and conifers, include boletes, amanitas, agarics, and many others. Endomycorrhizal fungi are associated with annuals and perennials, grasses, shrubs, and softwood trees. *Rhizobium* species (bacteria) are found almost exclusively on leguminous plants; *Frankia* species (actinomycetes) are found only in association with specific species of plants.

DISTRIBUTION Worldwide/ubiquitous.

GARDEN ACTIVITIES To plants: nitrogen-fixing, nutrient delivery; to soil: decomposers.

Springtails, Moss Piglets & Other Microarthropods

In addition to living in your garden, the remarkable tardigrades, also known as water bears and moss piglets, can survive temperatures of −459°F to 302°F, and live in glaciers, lava fields, and the deepest parts of the ocean.

An anxious inquiry once came my way from a homeowner who ordered 30 yards of compost and had it spread in his yard, only to ask for its removal because it was full of what appeared to be swarms of tiny hopping insects. They were not insects nor pests, I discerned from his description, but another of the small-size arthropods or "microarthropods," springtails. Springtails are also called snow fleas because they have the ability to thrive year-round, even in cold winters. Their most distinctive feature is a tail-like appendage found under their abdomen. When it releases, the springtail springs into the air. They do not have wings. They often appear gray, but closer inspection reveals that they are yellow or orange, and sometimes even purple or blue. They are common in gardens and will not harm plants, I assured him. In fact, they play an essential role in the decomposition of organic matter. We seldom notice them, but they can be present in large numbers in leaf litter, other organic materials, and, yes, compost. Although the caller was relieved to learn his compost was safe, the hired gardener was probably less pleased about having to move so much compost three times!

Springtails (Collembola)

Microarthropods include a truly diverse array of weird-looking creatures, and the main place to find them is in the thin but biologically active layer where soil and litter meet. Others are found in deeper soil layers; these are often thinner, wormlike in form, and with shorter legs (if they have them at all) than their counterparts in leaf litter.

Soil mites are the most common soil arthropods in all but arid and semi-arid environments (where ants play the leading role). About half of mite species are soil dwellers; some are predatory, but most are scavengers. I will single out oribatid mites, because they are the most abundant mite in the upper-soil layers. Sometimes called seed, moss, or beetle mites, these tiny scavengers are able to close their "shells" and resemble miniscule dark shiny beetles. Other mites dwell deep in the soil and little is known about them.

Pseudoscorpions are tiny, harmless arachnids unless you are a springtail—their preferred prey! They look like miniature scorpions, with two claws but no stinger. They often travel by hitchhiking on larger organisms such as daddy longlegs and insects (a relationship known as phoresy). You may spot the one illustrated here indoors, for "book scorpions" prey on the even-smaller pest of the glue in old books, booklice.

A myriad of other small arthropods may be found living in soil. Many species of small spiders inhabit leaf litter and the top layers of soil. Some minute centipedes and millipedes can also be found in soil, adults as well as juvenile forms of larger species. A distant relative of these is symphylans, occasional farm and garden pests. These dine mainly on decomposing matter, but if populations increase dramatically, they can turn to eating roots. Avoid them by not using partially decomposed compost (let it sit longer or, if you must use it, use it only on your garden soil's surface; do not mix it in).

Numerous species of insects are included with microarthropods. Even a seasoned beetle collector may be surprised at the diversity of tiny beetle species

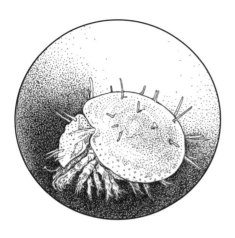

Oribatid mite (Oribatida)

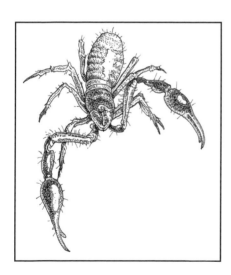

Book scorpion (Pseudoscorpiones)

found in soil. Ants, true bugs, crickets, and termites are also found in soil and leaf litter.

One miniscule soil organism that has captured people's attention is the tardigrade, also known as moss piglet or water bear. If you could see them (you would need a microscope), they really do look like tiny, plump bears. They are commonly found in soil, moss, and lawns, where they prey on protozoa and other organisms, and also feed on algae and fungi. They have the remarkable ability to shut down and reanimate years later.

Though usually found in water, a few species of rotifers are soil dwellers, or may be found in moss or lichen patches, navigating through a film of water in search of miniscule prey. These creatures are so small that they are often mistaken for single-celled protozoa, but in fact they are complex multicellular animals. If the soil dries out, rotifers may survive long periods of drought in the form of cysts, and reanimate (like the tardigrades) when conditions are suitable.

Nematodes are also part of this realm. While some species are the culprit in damage to plants, there are actually many beneficial nematodes. Those prey on species of pest nematodes—remember, it's a bug-eat-bug world. Others aid in decomposition.

Microarthropods and other small soil animals are usually visible (although sometimes barely so) to the naked eye, but to truly appreciate them requires magnification. A hand lens will do, but a sturdy small prism-style field microscope is ideal (I favor Brock Optical's Magiscope, magiscope.com). Speaking from long personal experience, looking much more closely at these tiny animals can provide endless entertainment for the inquisitive gardener!

The distribution of differently adapted species is a good reason to avoid disturbing soil layers. Many microarthropods are often short-lived but new generations are always coming on; many produce multiple generations over the course of a year. They inhabit all manner of available space and niches. In his wonderful book *Life in the Soil*, James Nardi points out—just as one of many good examples—that different springtail species may be separated by only thin layers of soil, and yet look and behave very differently.

We humans fret about disturbing nature in big ways (like paving over a vacant lot or cutting down forests), but we should also be concerned about disturbing nature on a smaller scale—in the soil layers in our gardens. Protecting the different communities of organisms in each layer of soil and litter, and their interactions, is an important reason why I recommend no-till gardening. Maintaining biodiversity in the soil food web supports aboveground biodiversity, and establishes a foundation for a thriving garden.

APPEARANCE Although many can be spotted with the naked eye, magnification is essential to get a good look.

POPULAR NAMES Mites, spiders, pseudoscorpions, and related groups; centipedes, millipedes, and related groups; crustaceans (small aquatic forms found in water features) and insects. Protura and Diplura are small soil arthropods related to springtails and insects. Non-arthropod soil organisms include tardigrades, rotifers, and nematodes.

SCIENTIFIC NAMES Chelicerata, Myriapoda, Crustacea, Tardigrada, Rotifera, Nematoda.

COMMON TAXA Most easily spotted microarthropods are oribatid and other soil mites, collembola, minute beetles, pseudoscorpions, and aquatic crustaceans. Minute centipedes, millipedes, and nematodes are also common.

DISTRIBUTION Worldwide. Many species of microarthropods and other small soil animals are not yet described.

GARDEN ACTIVITIES Varied. Some function as decomposers; many are predators of other small organisms.

The Roles of Roly-Polys, Potato Bugs & Other Macroarthropods

Soil dwellers include some fantastic-looking larvae. Antlion larvae, some-times called doodlebugs, are reputed to have been the model for the terrifying sand creature in the *Star Wars* movies.

By now, you may be wondering how there can be room for yet more life in the soil. We have explored earthworms, decomposers, nutrient facilita-tors, and the curious world of microarthropods. However, many larger, more visible and familiar arthropods are also found in soil and in leaf litter—these, we call the "macroarthropods."

Roly-polys and earwigs are familiar soil denizens. Many native bees and wasps build nests in the ground. In warmer weather, predaceous beetle lar-vae are abundant and active. The often-reviled Jerusalem cricket is found in western North America, ranging into Mexico and Central America. Other arthropods have larval stages that feed on roots; these are often specialists on specific plants.

Topsoil and leaf litter harbor some arthropods that spend their entire life in the soil, while others spend only a part of their life cycle there and are more easily spotted aboveground. Certain spiders, earwigs, and beetles spend the day hiding out of sight, only emerging at night. Many insects pupate in soil or leaf litter, spending winter safely tucked away and emerging as adults when spring returns.

Ground-nesting wasp (*Bembix* sp.)

Lifespans for larger soil arthropods vary. Some live a year or less. However, larval and pupal stages may take years in some species. For instance, some species of cicada nymphs live in the soil for up to 17 years, leading to periodic "booms" of up to 1.5 million adult cicadas per acre!

What are they up to? Soil macroarthropods most often function as detritivores, eating organic matter in various stages of decomposition. They facilitate soil structure as they tunnel through the soil, creating channels through which smaller organisms, water, air, and roots can travel. Still others, such as soldier fly larvae, are found just below the surface of decomposing organic matter, at the interface of soil and litter. A few are herbivorous and may eat roots, but in a garden with a diversity of plants, they're unlikely to have a significant impact.

Roly-polys (also called pillbugs) and their near relatives sowbugs are the only fully terrestrial crustaceans in North America; most common species are introduced. Crustaceans are distinguished by having gills, and therefore need moisture to survive. For this reason, we often find these creatures in humid, wet, or damp areas. Some are even fully aquatic.

How can you tell them apart? Roly-polys roll into a ball as a defense, to the everlasting delight of children. Sowbugs are flatter, have two short "tails," and are unable to roll up. Both, like other arthropods, must molt their exoskeleton to grow. They are unique in molting only one end at a time, with a few days elapsing before the second half is shed. The result can be creatures with differently colored halves.

Roly-polys and sowbugs mainly eat decaying and dead organic material, although gardeners (myself included) sometimes complain that they have a taste for strawberries and violets. In that eventuality, I recommend a judicious sprinkle of Sluggo Plus (iron phosphate, for slugs, plus spinosad, an organic pesticide), but only where damage is found. This will not only eliminate roly-polys, but also slugs, snails, and earwigs, yet won't affect other animals and the environment. Usually these creatures are harmless garden residents.

The common earwig in North America was introduced from Europe, probably hitch-hiking in shipping materials. They rarely fly, although many species have large membranous wings tucked under their short front wings ("elytra"); at the tip of the abdomen, they have formidable-looking pincers. One use for pincers is to carefully fold the flying wings under the elytra. The pincers are also used for defense against predators. A host of creatures dine on earwigs, including birds, certain mammals and amphibians, lizards, spiders, centipedes, and more.

Surprisingly, most earwigs are omnivores, detritivores, and even predators. Because they tend to be most active at night, it may take some detective work to identify their presence and any damage or benefit. Look under pots, a favorite hiding place. When present in large numbers, they can occasionally become significant pests in garden environments. Should that happen, they can be controlled with traps. A neighbor uses empty cat-food cans baited with a dab of soy sauce and some cooking oil; they come for the soy sauce and get mired in the oil.

But—good news—especially in trees, they may perform as predators, and they are known to dine voraciously on aphids. They will also consume the eggs of other insects, maggots, grubs, and armyworms. In fact, in Europe, "earwig homes" are sometimes suspended in trees to encourage them. Earwigs have also been found to function as beneficial predators in orchards here in North America. Should their food supply get low, however, they may turn to your garden plants. They favor tender young plants such as seedlings in the vegetable garden and like to climb into juicy, ripening fruit, such as peaches.

The poor potato bug, or Jerusalem cricket, suffers from an identity crisis. It only incidentally eats potatoes, preferring to dine on decaying vegetation and an occasional insect. Its formidable mandibles reveal that it is not a true bug (which have piercing, sucking mouthparts). Nor is it from Jerusalem, but from the western United States and Mexico, where it is known as "niño de la tierra"

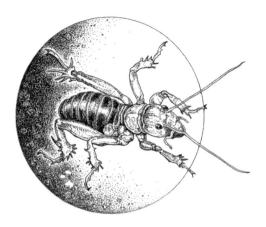

Jerusalem cricket (Stenopelmatus fuscus)

or "child of the earth." This large insect can deliver a painful bite when handled. In the garden, I sometimes unearth them in the compost or find them under pots and when digging in the rich soil of our garden beds. But they are infrequently seen, even where native, doing most of their feeding and moving around at night.

The aptly named antlions are drab-colored fliers that superficially resemble damselflies. The voracious little larvae, called doodlebugs, dig "death traps," funnel-shaped sand pits, where they trap and consume ants. Should you be lucky enough to have antlions in your garden—probably in a dry pathway—make sure water from sprinklers or hoses doesn't damage their lairs.

Any yard with areas of deep sand may host another small predator, the sand wasp (*Bembix*). Their front legs have rakelike spines for productive hole-digging. These holes function as nests rather than traps. Their favored prey are flies, which they paralyze and bring back to their nests for their larvae. Adults forage for nectar on flowers, particularly those in the aster family.

To encourage soil macroarthropods, let natural mulch accumulate under shrubs and trees, and disturb it as little as possible so they will have a place to live, eat, and reproduce. For plants such as roses and fruit trees that are prone to fungal diseases, spread a thick layer of new mulch on top of the existing litter as an alternative to raking up and disposing of decomposing leaves. As the autumn leaves fall in my shrub borders, I take pleasure in knowing that I am encouraging the next generation of allies that will greet me come spring.

APPEARANCE Macroarthropods are visible to the unaided eye, and vary in size, 0.13–2.5 in. (3–64 mm).

POPULAR NAMES Roly-polys, sow bugs, earwigs, Jerusalem crickets, and more.

SCIENTIFIC NAMES Phylum Arthropoda, subphyla: Chelicerata—spiders, scorpions, sea spiders, horseshoe crabs; Myriapoda—centipedes and millipedes; Crustacea—brine shrimp, barnacles, crabs, seed shrimp, and roly-polys; Hexapoda—insects and a few other small groups.

COMMON TAXA Roly-polys, earwigs, antlions, ants, crickets, ground beetles, and spiders are all commonly spotted soil-dwelling macroarthropods. Other species include larvae of various flies, bees, wasps, and many beetles, both larval and adult.

DISTRIBUTION Many groups are widely distributed. Certain soils have characteristic arthropod fauna.

GARDEN ACTIVITIES Primarily consume decomposing matter; some eat other bugs. Many are a food source for birds and other creatures.

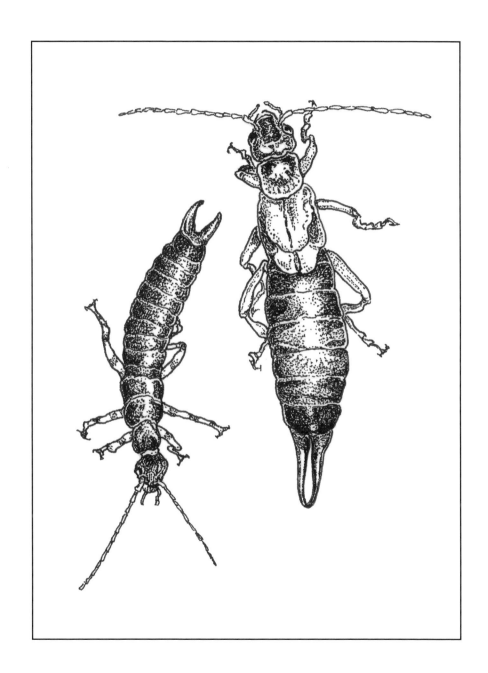

Left: Ringlegged earwig (*Euborellia annulipes*) Right: European earwig (*Forficula auricularia*)

2

ON THE WING

Bees, Butterflies & Other Flower Visitors

No doubt about it, blooming plants are busy places. Many insects seek sustenance from pollen, nectar, and even petals! A powdery or sticky substance composed of tiny grains (containing male genetic material), pollen offers nourishing carbohydrates and protein. Nectar, on the other hand, is mainly sugary water with traces of nutritious minerals. With few exceptions, our flower visitors benefit our gardens. Many are predators and parasites of pest insects, certainly a good thing. A great many other winged flower visitors are performing a vital task when they visit our blooming plants, pollination, and in that regard they are inarguably valuable allies.

According to one of my favorite books, *Evolution of the Insects* by David Grimaldi and Michael Engel, if we eliminated bees, ants, and termites from the planet, all terrestrial life would collapse. Whenever someone expresses a horror of insects, I pull this astonishing assertion from my bag of tricks. Without bees (and a few other six-legged pollinators), our diet would also become horribly bland. We would be living on potatoes, corn, wheat, and a few other crops. There would be no apples, pears, berries, citrus, zucchini, or tomatoes. No mint, thyme, oregano, or parsley. No chocolate, tea, or coffee! Most of the beautiful flowering plants that we take for granted would disappear.

Insects and flowering plants coevolved, leading to an explosion of species and the creation of multitudes of niches for other animals. Bees played a central role in increasing this biodiversity. The effectiveness of flower visitors as pollinators varies, but none can match the efficiency of bees; characteristic branched hairs on their bodies maximize the collection of pollen. One of the ways that bees can be distinguished from wasps (which they sometimes resemble) is that wasps lack branched hairs. Also, bees literally attract pollen due to an electromagnetic charge. Similar to the positive electric charge we humans experience when we scuff across a carpet, bees' bodies accumulate positive charges as they forage. Pollen is negatively charged. When the two meet (bees' branched hairs and pollen; we reach the other side of the room and touch a doorknob), there is a brief "static electricity" connection. So far, bees are the only insects found to carry this electromagnetic charge, although it is possible other flying insects and even hummingbirds experience this phenomenon; investigations are ongoing.

We begin this chapter, then, by exploring the diversity of the all-important bees. Ideally, you would like many species to become regular pollinators in your garden. They can be loosely grouped by type according to their dominant social behavior (three essays are ahead). Knowing what they eat is only half the equation, for they also need a home base. For an abundance of native bees, remember that most are ground-nesters and require bare soil, while others nest in hollow stems and twigs.

As for the butterflies, while some of the hairier ones can be effective pollinators, many perch high above the blooms where they gather nectar, collecting little pollen (including our beloved monarch butterflies). Moths are far better pollinators. Even stinging (aculeate) wasps must be included here, as they regularly visit flowers for sustenance as adults. Although their shiny, usually hairless bodies make them much less effective as pollinators, they collect many pest insects to feed to their larvae. We will explore the parasitic wasps in another chapter, as many don't require nectar as adults.

We would like to clearly differentiate between hunting wasps and parasitic wasps, but nature being what it is, the boundaries are sometimes a bit vague. Hunting wasps are not always larger, nor are parasitic wasps always tiny. Life history is a better indicator. For our purposes here, we define the hunting wasps as capable of carrying off their prey to their nest. Some are specialists, for example, the colorful wasps known as tarantula hawks. Specialists can be more effective at controlling garden pests than generalists, which are not as discriminating in their tastes and may even attack desirable garden insects. If you would like to further explore the subtleties of the differences between hunting and parasitic wasps, I highly recommend Eric Grissell's fascinating book *Bees, Wasps, and Ants: The Indispensable Role of Hymenoptera in Gardens*.

Beetles too sometimes visit flowers, as do some bats, but I will cover them in other chapters. Many garden flies are flower visitors; here we will highlight two of the most effective natural enemies of pest insects (bonus: they are also pollinators!), the hover flies (also known as syrphid or flower flies) and tachinid flies. They are not the only ones, of course. You will meet other flower visitors throughout this book.

One thing is certain: an abundance of the right flowers will bring in an abundance of winged garden allies. Fortunately, many of the same flowers that attract pollinators attract beneficial insects that attack pests. Choose open flowers with readily available nectar and pollen, and minimize flowers with crowded petals. Strive for a long season of bloom, offering a range of shapes, colors, and sizes. Plants in the aster, parsley, mint, mustard, and rose families are among your best bets. Annual native wildflowers are easy to grow and attract all sorts of flower visitors. You might be surprised to learn that some plants with inconspicuous flowers, such as willows and grasses, are also worth adding to your yard, as they are among the earliest spring flowers to bloom.

A Solitary Existence

Solitary bees produce fewer offspring than almost any other insect. It is not uncommon for some species to produce fewer than 25 during their entire life cycle.

Solitary native bees, the first category of bees we will explore here, comprise a very large group and contribute much of the bee diversity of temperate zones. They are indeed loners, although occasionally they live in groups (aggregations). The key is that they do not help each other in reproduction. Instead, unlike honey bees, each solitary female bee is reproductive and builds her own nest—there are no worker bees—and provisions it with the food necessary for her larvae to develop. Nor do solitary bees produce honey or wax. Most are either stingless or highly unlikely to sting. The stinger (in all stinging insects) is a modified ovipositor (a tubular organ that females use to deposit eggs); thus, only female bees can sting. However, solitary bees are not aggressive and those that can sting only do so in self-defense. Some species that nest in aggregations, including orchard and alfalfa bees, are commercially managed for pollination.

In their annual life cycle, generally only a short time is spent as an egg, larva, and adult, as compared with the pupal stage. Where the ground freezes in winter, solitary bees tend to have short adult life spans. Some species emerge at the first hint of warmth, and rely on early-flowering plants for sustenance. They

Solitary halictid bee (Halictidae)

may specialize in just a few or, more rarely, a single species of flowering plants ("oligolectic"), or have a broader, more generalist diet ("polylectic").

The oligolectic specialists, which are the majority of solitary bees, are found flying around only during the blooming season of their preferred plant species. They are often far less discriminating when seeking nectar (sugar) than pollen (protein).

All bees, though, require both sugar and protein to survive and thrive. With a few rare exceptions, only females gather nectar and pollen. A few carry pollen internally, but most bees have external structures for carrying pollen, tiny baskets or brushes. Called scopae, these are often on their hind legs. In the solitary leafcutting bees (family Megachilidae), pollen is carried on the underside of the abdomen (the venter), a distinguishing characteristic for that family.

The majority of solitary bees are ground-nesting. Among these are plasterer bees (family Colletidae); these characteristically line their nests with a cellophanelike material. They are relatively hairless and may be mistaken for wasps, especially as they carry pollen internally. Mining bees (family Andrenidae) are also solitary ground-dwellers. Some species may use a common entrance to the nest, but each female provisions her own offspring. The above-described pollen-gathering baskets (scopae) are usually present the entire length of mining bees' hind legs. Many are specialists on specific plants. Mellitidae is a very small family of mostly ground-nesting solitary bees with restricted distribution. They can be identified by their shaggy pollen-carrying baskets (scopae). As all of the ground-dwelling bees prefer bare ground, and are often found nesting in sun-baked banks, leave some areas in your yard mulch-free for their benefit. The space between your paths and garden beds is ideal nesting habitat.

Other solitary bees seek out hollow stems or cavities in wood. Mason, leafcutting, and carder bees (all family Megachilidae) mainly nest alone. A few may nest in aggregations, but it's the exception. Mason bees use mud as a nesting

Leafcutting bee (Megachilidae)

material, while leafcutting bees build nests lined with leaves and sometimes, charmingly, petals. Carder bees line their nests with animal or plant fibers.

Populations of many species of wood-nesting bees can be accommodated with a simple homemade "hotel" fashioned from a bundle of hollow reeds or bamboo sections, nestled into a coffee can or placed in a sheltered area. Place these friendly invitations to join the garden allies team in a protected spot, out of rain and wind. I recommend the Xerces Society book, *Attracting Native Pollinators,* for detailed guidelines.

But wait! There are more! Sweat bees (family Halictidae) and honey bees, bumble bees, and carpenter bees (family Apidae) exhibit a variety of social behaviors, from purely solitary to the teeming colonies of honey bees. We'll delve into them next.

APPEARANCE Adults vary in length, 0.1–1+ in. (3–25 mm). They may be brown, black, or sometimes metallic green or blue; some are striped yellow, white, orange, or red. Carpenter bees are larger than most, around the size of a bumble bee (0.75–1 in.).

POPULAR NAMES Plasterer bees (Colletidae); mason, leafcutting, and carder bees (Megachilidae); mining or digger bees (Andrenidae); Melittidae (no common name); sweat bees (Halictidae). Honey bees, bumble bees, and carpenter bees (Apidae) include a few solitary species.

SCIENTIFIC NAMES Order Hymenoptera, superfamily Apoidea (bees). Six of the seven families of bees are found in North America; all include solitary species.

COMMON TAXA Megachilid bees such as the native blue orchard bee (*Osmia lignaria*) and the non-native alfalfa bee (*Megachile rotundata*).

DISTRIBUTION Worldwide, about 20,000 species of bees. About 4000 native to America north of Mexico: 160 Colletidae, >630 Megachilidae, >1200 Andrenidae, 32 Melittidae, 520 Halictidae, about 1000 Apidae.

GARDEN ACTIVITIES Valuable pollinators; solitary bees seek pollen and nectar in both larval and adult stages.

BEES

Solitary to Social Bees

In desert habitats, some bee species visit flowers only at dawn and dusk, avoiding the heat of the day ("crepuscular" is the term for this habit). Flying in low light, they often have unusually large eyes.

Strolling through my garden early in the morning, dew heavy on the flowers, I sometimes pause to gently pet a velvety sleeping bumble bee. Although it may seem a strange activity to some, for me it embodies the ephemeral magic of the garden at dawn. I can touch the slumbering bumble bee because of an unusual behavior—some male apid bees may sleep on vegetation, clasping stems or petals with their mandibles and spending the night at rest. Bumble bees are the largest of our native bees and the most familiar "in between," hard-to-categorize-neatly bees. Here we explore bees that may exhibit a variety of solitary behaviors (such as provisioning their own nests) or social behaviors, depending on species. Some of these, such as bumble bees, live in annual social colonies, but only the honey bees (in the next essay) establish perennial colonies.

Bumble bees are apids (family Apidae), a group with the broadest range of social behaviors of any bee family, exhibiting the full gamut from solitary (see previous essay) to truly social (eusocial) (see next essay). Other apids include squash bees, long-horned bees, carpenter bees, honey bees, and several lesser-known groups. They are closely followed in social diversity by bees in the family Halictidae.

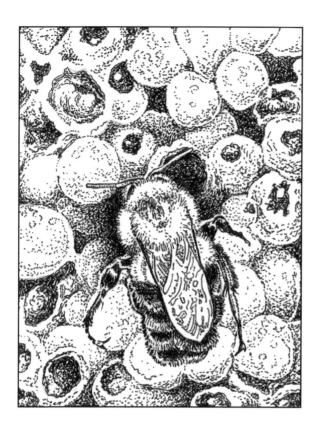

Bumble bee (*Bombus* sp.) in nest

Between solitary and advanced bees there are many social behavior variations, depending on species. Members of the same generation may cooperate in brood care to varying degrees (this is called "quasisocial" behavior). Communal bees may share a common nest, but each female provisions her own cells with food for larvae. In other species, there is division of reproduction, with some laying eggs and their sisters of the same generation acting as workers (these are "semisocial" bees). In all these variants, the bees are annual nesters.

Bumble bees establish their annual colonies in old rodent burrows and other ready-made homes. Only the mated new queens overwinter. Like honey bees, they are generalist feeders, and great pollinators of solanaceaous plants (tomatoes, peppers, eggplants, potatoes, petunias, and more). They dislodge pollen by vibration, and are more active in low-light levels and cool weather than honey bees. Soon after dawn, bumble bees awaken and another day's activity begins.

Bumble bees are "corbiculate" bees, named for the "corbula" or pollen baskets on their hind lower legs (tibias). In this, they are related to honey bees, orchid bees, and stingless bees (the last two occur only south of the United States border). Corbiculate bees are the only bees that ever exhibit advanced social behavior and, as noted, they don't do so all the time.

Some bees exhibit an opportunistic behavior called "kleptoparasitism." A good example is the slender, hairless cuckoo bee (these resemble stinging wasps, though they are not aggressive in nest defense). Using the same strategy employed by their bird namesakes, cuckoo bees lay their eggs in other bees' nests, waiting and watching till the other bee heads out to forage before sneaking in. When the cuckoo bee larva hatches, it kills the host egg or larva, if the egg-laying adult hasn't, then eats the host larva's pollen stash. Cuckoo bees aren't the only ones. The entire apid subfamily Nomidinae behaves in this way, as do some other apid, halicitid, and megachilid bees. Basically they outcompete the rightful larva. There is a fascinating chapter on this curious behavior in *The Bees in Your Backyard* by Joseph S. Wilson and Olivia Messinger Carril.

Some of my favorite garden bees are the beautiful halictid bees. The female is uniformly bright metallic green; the male sports a handsome black and yellow striped abdomen. Halictids are also called sweat bees, as some species are attracted to perspiration. Some nest in decaying wood, but most are soil dwellers, digging their own nest. Halictid bees are unusual in that social behavior may vary even within a single species. They are important pollinators of a wide range of plants, from fruit trees to wildflowers.

Carpenter bee (*Xylocopa* sp.) stealing nectar

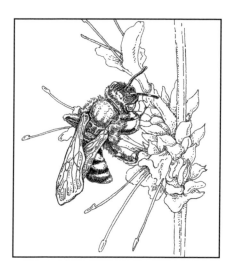

Green metallic sweat bee (*Agapostemon* sp.)

Besides bumble bees, other apids include carpenter bees, both small and large. Large carpenter bees are familiar to most gardeners; they look like shiny black bumble bees. Most are solitary species, but in some species mothers and daughters may occupy the same nest. Some may divide labor. Even solitary species are sometimes found in groups (aggregations). Tiny solitary metallic black carpenter bees nest in the stems of woody plants. Carpenter bees often behave as nectar robbers, cutting a slit at the base of a flower and taking the nectar without providing pollination—an amazing sight if you ever get to witness it!

Digger and squash bees include many specialist pollinators. In gardens, the solitary squash bees are commonly found wherever squash, pumpkins, and other cucurbits are grown. They arrive early in the morning, and some are even able to fly in the dark—recognize them by their large eyes. Most are soil-nesting, sometimes in aggregations. The commonly encountered long-horned bees (*Melissodes* spp.) bees are also in this group.

As for diet, these in-between species are like other bees, foraging for nectar and pollen. A few collect floral oils (mixing the oil with pollen as food for their developing young, unlike most bees that mix pollen with nectar). While I have divided the bees in terms of social behavior, those who study native bees often categorize them according to tongue length, which determines which flowers they visit. For instance, the apids are long-tongued and halictids are short-tongued. Any garden that has a diversity of flower forms and sizes will attract a diversity of bees. Do you need another reason to plant a variety of beautiful flowers?

APPEARANCE Adults, 0.25–1+ in. (6–25 mm), brown, black, or sometimes metallic green or blue; may be striped yellow, white, orange or red. Baskets (scopae) on legs carry pollen.

POPULAR NAMES Halictid bees include the sweat bees and green metallic bees. Apids include long-horned, squash, and bumble bees. Most common groups of apid bees range from solitary to social bees.

SCIENTIFIC NAMES Order Hymenoptera, superfamily Apoidea (bees), families Halictidae and Apidae.

COMMON TAXA Halictidae include *Agapostemon* species and Apidae include *Melissodes*, *Peponapis*, and *Bombus* species.

DISTRIBUTION Widely found in America north of Mexico. 520 species of Halictidae; about 1000 species of Apidae.

GARDEN ACTIVITIES Pollinators of a broad range of garden flowers, vegetable crops, fruit trees, and native plants.

Busy Honey Bees, Truly Social Insects

Only female bees sting. The honey bee has a barbed stinger, and the stinger and the venom sack remain in the unfortunate human's skin, while the bee dies.

In the late 1600s, a Dutch biologist named Johannes Swammerdam meticulously dissected a honey bee with homemade tools and a microscope so primitive it would be difficult to recognize it as such today. In doing so, he discovered that the "king" bee possessed ovaries and was in fact a "queen." He also demonstrated that insects had social systems as complex as any other animal. Another very significant discovery is credited to him; he was the first to document the process of metamorphosis. He showed that caterpillars and other larvae were not completely different animals from the adult forms. His drawings are still unparalleled; these and his remarkable discoveries can be explored online (janswammerdam.org).

Honey bees have long enthralled humankind. They are an enduring motif in cultures around the world. In Greek mythology, priestesses for the goddesses Demeter and Artemis were known as the "Melissae" (Greek for bees). Bees are featured in the creation myth of the San people of the Kalahari Desert, and appear as Freemason, Hindu, and Christian symbols. The Mayans were known to worship a bee god, and the Egyptians revered bees as a symbol of the sun

Honey bee larva, worker, queen, and drone

god Ra. Honey bees are regarded as hard-working, industrious, and cooperative. Their social structure is universally admired and emulated.

Our desire for honey and wax led us to domesticate honey bees. Indeed, historians note that doing so was a significant milestone in the rise of early civilizations. Honey bees' contribution as pollinators cannot be overstated. By some estimates, well over 30 percent of modern crops depend on them. In this country, hives are often moved long distances, literally "rented" to provide pollination services for crops. One-third to one-half of all the hives in the United States, for instance, are trucked into California's Central Valley to pollinate the world's largest almond orchards every spring. The "pollination circuit" next goes to New York for apples, to Maine for blueberries, to Massachusetts for cranberries, and then back to the South for winter. Farmers also need honey bee pollinating services for crops of cotton, soybeans, and buckwheat, and orchard fruits such as apples, plums, and pears.

Movement of hives is undoubtedly a factor in the many health problems honey bees face today, as it contributes to the spread of diseases and parasites. Another threat may be the corn syrup that is often fed to bees as an overwintering nutrient in commercial operations; it's a bit like trying to live on fast food all winter. Pesticides, especially neonicotinoids, are widely implicated in the depletion of bee populations. You may see neonicotinoid-treated plants for sale labeled as such (which allows concerned gardeners to avoid them).

Of course, humans also appreciate honey. We have sometimes gone to great lengths to acquire it—Himalayan honey hunters, for instance, gather it while clinging to sheer cliffs, an ancient practice that continues today. The sweet flavor, which varies according to a colony's diet, is delicious. But we also value its various health benefits, including its antioxidant content. Honey, for instance, soothes a sore throat and has antifungal and antibacterial properties. You can even dab it on a cut or abrasion. Raw honey is preferable to pasteurized, as the heating process diminishes the nutrients, although it also kills unwanted yeast and prevents crystallization.

Other bee products are also prized by humans. With the advent of paraffin (which is a petroleum product), we take candles for granted, but wax from bees was once the principal source of household candles. We have learned to harvest propolis, a mixture of resins gathered and used by bees as a sealant; we value it for its medicinal and antibacterial properties. Royal jelly, a glandular secretion fed to larvae and queen bees, has a long history as a health and beauty treatment. Allergy sufferers sometimes eat regionally produced honey and pollen to better tolerate the effects of local blooms. Even the bees themselves have been enlisted. A long-time folk remedy, now supported by science, is bee-sting therapy to reduce the pain and inflammation of arthritis.

In the wild, bees live in colonies in trees and other hollows, while domesticated bees live in hives. You may be surprised to learn that honey bees are not native to North America, but originated in the Mediterranean. There was

Honey bee worker with pollen ball

Honey bee worker at its nest

no American Indian word for honey bees—and they became known as "the white man's fly"!

Honey bees exhibit advanced social behavior, termed "eusocial." There are three key traits: a division of reproductive labor (only the queens reproduce); bees cooperate in caring for immatures; and there exists an overlap of generations within colonies (daughters are workers). In advanced eusocial bees, there are also physical differences between queen and workers. Honey bees live in perennial colonies (and differ in this from bumble bees, which have only annual colonies), with a generally solitary queen bee that lays all the eggs. Unfertilized eggs become drones, while fertilized eggs become workers. A fertilized egg laid in a special cell is supplied with only royal jelly as larval food and becomes a queen. The queen lays hundreds of eggs a day, especially in spring, as worker bees only live a few weeks and must be replaced. Workers divide labor according to life stages, beginning their adult stage with the care of larvae and queen. A few days later, they switch to hive maintenance tasks such as waxwork and nectar processing. At about three weeks, most worker bees become foragers, while others take on roles in defense, or disposal of dead bees.

Honey bees collect pollen, a valuable source of protein for larvae, in their back leg "pollen baskets" (corbiculae). They are able to communicate the direction and distance from the hive of good sources of nectar and pollen with a delightful "waggle dance." Your garden could very well be on their list to visit! Honey bees feed on a wide variety of flowering plants but generally concentrate on a single plant species at a time, making possible the diverse flavors of honey, such as the clear flavor of clover, the floral note of orange blossom, and the deep richness of chestnut. Mixed sources yield "wildflower honey."

No summer garden is complete without the pleasant hum of honey bees buzzing about the flowers. Encourage their presence by growing a diversity of flowering plants, planting bee-attracting plants in clumps, and eliminating pesticides from your garden management. Find good plant lists online (the Xerces

Society has lists for all regions of the country) or ask your local nursery or garden center what they recommend. Last but not least, relax on mowing and weeding a bit. Honey bees do love dandelions and clover.

APPEARANCE Dark and light strains, but adults are generally golden, with brown and golden yellow striped abdomens.

POPULAR NAMES Honey bee, European honey bee, western honey bee.

SCIENTIFIC NAMES Order Hymenoptera (ants, wasps, and bees), superfamily Apoidea (bees), family Apidae.

COMMON TAXA Over 20 strains (subspecies) of *Apis mellifera*, adapted to different conditions. Hybrid strains also exist, including the Africanized honey bee. Strains and hybrid strains may exhibit great variability in color, behavior, aggressiveness, and other characteristics.

DISTRIBUTION Worldwide, except Antarctica.

GARDEN ACTIVITIES Our most active pollinators!

The Marvel of Moths

How can moths be garden allies? Simple: Over 95 percent of terrestrial birds feed their babies insects, especially caterpillars, and most caterpillars are moth species.

Butterfly gardens continue to grow in popularity. Gardeners who once looked askance at including food plants for caterpillars now eagerly seek out information on larval host plants for butterflies. Moths, however, remain overlooked, even though the vast majority (over 95 percent) of lepidopterans (the order that includes butterflies and moths) are moths. True, some moths are among our worst horticultural and agricultural pests. However, many garden moths are well-camouflaged, both as adults and larvae, and damage can be difficult to detect. In fact, most moths are harmless or offer benefits to our plants and gardens. Many play an important role in pollination.

Gardeners who plan for biodiversity by including native plants in their landscapes, and refraining from pesticide use, unintentionally develop "moth gardens." One result is that birds arrive. Many species of moth larvae are critical bird food, especially when bird parents are seeking high-protein nutrition for their young. Thus the populations of many species of moths are kept under control. A moth-rich garden becomes a bird-rich garden.

Some introduced moths are subjects of debate even when there is a lack of documented plant damage. A good example is the accidentally introduced light brown apple moth (*Epiphyas postvittana*, referred to as "LBAM"), which feasts on dozens of valued plant species in its native environment, not only apple trees. The LBAM is considered a potential threat, as it is as yet

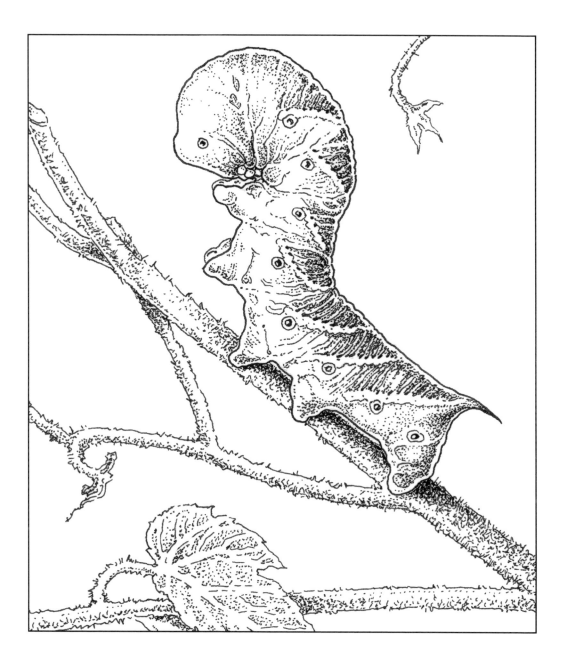

Tomato hornworm (*Manduca quinquemaculata*)

unknown if the species will reach a "tipping point" and someday become a significant problem.

Most moth species have a narrower range of larval host plants than does the LBAM. Some are even restricted to a single genus, or even to only a few species of plants. One example is the artichoke plume moth (*Platyptilia carduidactyla*), which attacks a number of thistle genera in addition to artichokes.

The spectacular ceanothus silk moth (*Hyalophora euryalus*) is a West Coast species whose caterpillars most gardeners would be pleased to encourage in their shrub border. Deceptively named, this moth feeds on a variety of other woody plants in addition to ceanothus, including willow, alder, and more. One year I led nature walks in Sonoma County near where I live. Regardless of itinerary, our path invariably passed near a certain coffeeberry shrub festooned with a dozen or so huge, beautiful, silvery beige cocoons. This dramatic sight always stopped everyone in their tracks. My own garden now includes coffeeberry, in addition to mountain mahogany, gooseberry, and manzanita, all larval host native plants for this beautiful moth. I am still waiting for some to show up so I can enjoy the beautiful adults!

Moths are not always easily differentiated from butterflies. With few exceptions, butterflies are active during the day and moths during the evening and night hours. (The beautiful, aptly named bumble bee moth, which can be seen dining on nectar in bright sunlight, is one of the rare exceptions.) Butterflies are usually more brightly colored than moths, but day-flying moths may closely resemble butterflies. Moth wings, like those of butterflies, are covered in scales and are held tented or flat at rest, whereas butterflies fold their wings vertically over their bodies. But the foolproof way to tell them apart is to get a look at the antennae. Butterfly antennae end in tiny knobs, while moth antennae are usually thread- or featherlike.

All that said, as often happens, there are "in-between" species. The common stout-bodied skippers resemble moths, but they are a link between butterflies

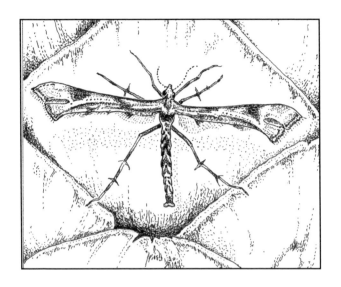

Artichoke plume moth (*Platyptilia carduidactyla*)

and moths. Look closely. The skipper holds its wings differently than either relative, and has characteristic hooks at the ends of its antennae. And there are mimics, moths that look like other insects. For instance, many clear-wing moths are day-flying wasp mimics (and however beautiful, are known to gardeners as wood-boring pests).

Because we gardeners tend to notice more things in the light of day, we are more familiar with the caterpillars of moths than with the nocturnal adult forms. The flying adult phase may only live a week or two; larvae endure longer. Some moth species have several generations annually, others only one.

Did you know that the commonly seen striped "woolly bear" caterpillar is the larva of a tiger moth? It feeds mostly on weedy plants. A popular bit of folklore is that the size of the woolly bear's central stripe is an indicator of the severity of the coming winter. The colorful wings (for a moth) warn predators

Bumblebee moth (*Hemaris diffinis*)

like birds that they are not good to eat (in fact, their body fluids contain neurotoxins). As adults, these moths are generalists and feed on the nectar of many plants. In a word, from the point of view of gardeners, this species in all phases of its life cycle is harmless.

The destructive tomato hornworm is another story. Most gardeners are quick to kill it on sight, assuming they can even spot it, as its green color allows it to blend in on the plant. It turns out to be the larva of the beautiful, heavy-bodied sphinx or hawk moth (*Manduca quinquemaculata*) or, sometimes, its close relative the tobacco hornworm (*Manduca sexta*). In its adult form, it can be seen at dusk as it hurtles around visiting flowers; it is reputed to be able to pollinate more than 200 flowers in less than seven minutes—impressive! That creates a bit of a conundrum for gardeners. If you have only one or two, no problem, but if your tomato plants are infested, you'll want to step in. Instead of killing them, you might pass the unwanted caterpillars along to a teacher for a great classroom activity. Or sacrifice one tomato plant; leave a few hornworms on one plant and swathe it in cheesecloth, pinning all exits with clothespins, until they enter the pupal phase (they do this in the soil). Then unwrap the plant and await the emergence of these productive pollinators. Note that hornworms are sometimes parasitized by tiny wasps; learn more in Chapter 3.

It is interesting, and sobering, to note that of all our hardwood trees of forest, park, and garden, it is various oak species that seem to be the most vulnerable to moth damage. In many parts of Canada and the United States, the voracious European gypsy moth (*Lymantria dispar*) preferentially feeds on oak trees, and often causes significant damage. Tent caterpillars (*Malacosoma* sp.) are easily spotted aggregating in their silken tents along tree branches in mid- to late summer; they are capable of completely denuding a host tree. In warmer regions, the fruit tree leafroller (*Archips argyrospila*) may also strip oaks of their leaves. In my home state of California, in some years, the native oakworm (*Phryganidia californica*) completely defoliates the native live oaks that are their preferred oak hosts. Thankfully the trees usually recover in the same growing season. One moth specialist I have spoken to suggests that since oak leaves break down so slowly, outbreaks of native caterpillars may be fertilizing the oaks. He is quick to point out that his intriguing speculation remains untested. Let's remember that although 5000 species of insects (including dozens of moth species) may be found in native oak trees, we rarely notice the damage. Instead, we notice the birds arriving for a feast.

At the central California botanic garden where I once worked, Western tussock moths (*Orgyia vetusta*) would drop from the oak trees on silken lines in spring, piquing children's curiosity. I always enjoyed relating the description I first heard in a long-ago entomology course, about the flightless female adults, "a hairy bag of eggs," as I watched their young faces light up with delight at yet another of nature's wonders.

So, are moths garden allies? We know that caterpillars are valuable bird food, but what about the adult moths? It turns out that they are effective pollinators. Remember that they are out and about at dusk, when light is low. I know of a woman in Maine who grows moonflower vines over her terrace for the simple pleasure of sitting nearby in the cooler evening air to enjoy the visiting moths, watching as they unroll their extra-long tongues down into the trumpet-shaped blooms for nectar. She has planned for a special joy, at least for the summer months. If you would like to offer tempting beacons to pollinating moths, grow some white or light-colored flowers, especially if they are fragrant. Native members of the lily family are attractive to moths, but they're also drawn to nicotiana (flowering tobacco), yucca, gardenias, and many native flowers.

APPEARANCE Moths are often dull-colored and camouflaged, especially nocturnal species. Day fliers may be brightly colored and resemble butterflies. Moth antennae are usually feathery or, if threadlike, with no ball on the end.

POPULAR NAMES Moths.

SCIENTIFIC NAMES Order Lepidoptera; all families that are not butterflies are moths.

COMMON TAXA Tiger moths; larvae are woolly bear caterpillars (family Erebidae, subfamily Arctiinae). Geometrid moths; larvae are inchworms and loopers (Geometridae). Owlet moths; larvae are earworms, armyworms, and cutworms (Noctuidae). Sphinx moths; larvae are hornworms (Sphingidae). Silk moths; larvae have no group common name (Saturniidae). Many other families of moths are found in gardens.

DISTRIBUTION About 12,000 species in America north of Mexico. Worldwide, about 160,000 species. Many remain undescribed.

GARDEN ACTIVITIES Moth caterpillars dine on foliage of trees, shrubs, herbaceous plants, vegetable crops, and fungi. Adults can act as pollinators; a few are parasitic. Many are detritivores (for example, clothes moths).

Butterflies

Monarch butterflies, the poster child for pollinator gardens across North America, are not good pollinators! Perched high on flowers, they don't readily pick up and move pollen. They are, however, important indicators of ecosystem health.

Butterflies are well-loved even by gardeners who profess profound dislike for other insects. Mythologies and folk stories, art, music, and literature are rich with butterfly imagery. "Butterfly gardens" full of flowering plants provide nectar for these favorite insects in both private and public landscapes. Their lifespans, typically one or two generations annually in temperate climates, allow us ample opportunity to admire them.

Ask where the "caterpillar garden" is located, however, and you will likely be met with incredulity. Undoubtedly, the vision of tattered foliage draped on bare branches gives pause to many a well-meaning butterfly gardener when mulling over the inclusion of a banquet for voracious leaf-eating insects. Yet without caterpillars, there are no butterflies to enjoy the sweet nectar we provide. A carefully planned garden can include caterpillar habitat and still be beautiful. While it is true that many host plant species suffer damage, this is often insignificant or at least only temporary. Site your butterfly larval host plants in an inconspicuous spot or a mixed bed, where they will blend in.

Caterpillar garden design is dictated by caterpillars' generally finicky nature when it comes to diet. Most caterpillars are tightly coupled to a narrow range

Anise swallowtail butterfly (*Papilio zelicaon*)

of host plant species, starving to death rather than accepting a substitute (this is true for most species of herbivorous insects). This now well-known obligate ecological relationship was the foundation for a seminal 1964 scientific paper on evolution by biologists Paul Ehrlich and Peter Raven, in which they documented the association of certain butterfly and plant species and related it to the toxic secondary compounds produced by plants. They popularized a now familiar word, "coevolution," defined as "a reciprocal evolutionary exchange in two interacting species." They then asserted that plant diversity was inextricably connected to herbivorous insect diversity.

Certain insects and plants are more tightly coupled than others. For instance, monarch caterpillars feed exclusively on milkweed, and pipevine swallowtail caterpillars on Dutchman's pipevine. There is an advantage to their specialized diet, as compounds in the plants make both the caterpillar and the butterfly that follows taste bad to birds and other predators. Many predators learn to avoid them, while other species have evolved strategies that allow them to eat the toxic treat.

The bright colors of the caterpillar and, later, the wings serve as a warning. Species that are not ill-tasting or toxic mimic their colorful appearance, a bid for self-preservation. Other examples of close associations include fritillaries, which use members of the violet and passion vine families as larval hosts. Skippers favor grasses. The caterpillar of the California dogface, that state's butterfly, is found only on the beautiful, fragrant, purple-flowered California false indigo shrub, *Amorpha californica*. These relationships are complex and evolve over time. Plants become more toxic as caterpillars adapt to eating them. In turn, the caterpillars' predators adapt to be able to eat them. Coevolution at work!

Other caterpillars enjoy a somewhat broader diet. The sulfurous mustard family, for example, serves as the larval host to a number of butterfly species that can feed on many species in this large family, and some plant species, such as oaks and willows, host several species of butterflies.

In some cases, the caterpillar that precedes the butterfly is too great a pest to tolerate. The cabbage white butterfly, an introduced species, is a prime example. Its little green caterpillar feeds on many of our favorite garden brassicas, from broccoli, cabbage, and kale to nasturtiums and wallflowers. Prevention is your best strategy; protect vulnerable plants with netting early in the growing season. If the caterpillars appear and start nibbling holes in your plants' leaves, pick them off and dispose of them or try a distasteful garlic or hot-pepper spray (avoid toxic sprays, which endanger the soil food web as discussed in the first chapter). If you use *Bacillus thuringiensis*, a bacterium that only kills caterpillars, use it only on infested plants and do not spray on a windy day. Note that it is only effective on early larval stages (instars). Natural predators include house sparrows, goldfinches, and skylarks. Adding nesting boxes that attract these birds may help reduce the caterpillar population.

Generally speaking, most native caterpillars prefer native plants—coevolution at work. But inevitably there are exceptions to the rule. Some species appear to have partially or even largely switched to introduced plant hosts. For

Monarch caterpillar (*Danaus plexippus*)

example, the native anise swallowtail is frequently found on an introduced plant, Mediterranean fennel, common along roadsides and in empty lots throughout the western United States. (When it was proposed to eliminate fennel from San Francisco some years ago, butterfly lovers protested that it would reduce habitat for the swallowtail.) Adults dine on nectar from flowers, but also sometimes on rotting fruit. Some butterflies obtain minerals from mud puddles and excreta of other animals.

The Xerces Society (which I mentioned in the bee essays), a national insect conservation organization named for the now extinct Xerces blue butterfly, is a great source of advocacy for and information about butterflies. That extinct butterfly is emblematic of the many butterfly species that are in danger of a similar fate. Lepidopterist Art Shapiro has studied butterfly populations for over 30 years, and has documented an alarming general decline in the numbers of many species. At least some this decline can be attributed to habitat fragmentation and loss; climate change is also implicated.

California dogface butterfly (*Zerene eurydice*)

Public awareness of the plight of these charismatic insects is increasing. Several national organizations are involved in butterfly conservation, and regional efforts abound. The Xerces Society were the ones to kick off a nation-wide milkweed planting campaign to increase habitat for monarchs. Closer to my California home, the Green Hairstreak Project in San Francisco seeks to increase urban habitat corridors for butterflies, and the green roof of the California Academy of Sciences in Golden Gate Park includes larval host and nectar plants for the public to view, enjoy, and draw inspiration from. In my own neighborhood, an enchanting butterfly sanctuary created by pipevine swallowtail enthusiast Louise Hallberg provides habitat for over 40 species of butterflies (and caterpillars) and attracts thousands of visitors annually. Around the world, similar efforts are taking place, including, I am sure, some-where near wherever you live and garden.

APPEARANCE Eggs spherical or oval (ovate), laid singly or in clusters on host plants. Caterpillars have three pairs of true legs and five or six pairs of "prolegs." Pupae vary in appearance. Adults have two pairs of scale-covered wings, often brightly colored. Antennae are threadlike with a ball or club on the end.

POPULAR NAMES Butterflies.

SCIENTIFIC NAMES Order Lepidoptera, families include brush-footed (Nymphalidae); whites and sulphurs (Pieridae); swallowtails (Papilionidae); gossamer-winged, including blues, coppers, and hairstreaks (Lycaenidae); metalmarks (Riodinidae); and skippers (Hesperiidae).

COMMON TAXA Monarchs (*Danaus plexippus*), painted lady (*Vanessa cardui*), cabbage white (*Pieris rapae*).

DISTRIBUTION Painted lady and monarch butterflies are almost global in distribution, and known for their spectacular migratory behavior. Most other species have more restricted ranges.

GARDEN ACTIVITIES Many species are plant host–specific in larval phases. Caterpillars eat leaves and adult butterflies are nectar feeders.

Tarantula Hawks, Mud Daubers, Paper Wasps & Hornets

Don't like wasps? Good news! Most mind their own business of hunting—and most pest insects on the planet have at least one wasp species that preys on it.

M ost of us gardeners are enthusiastic in welcoming beneficial insects to our gardens. Lady beetles, lacewings, and even native bees of all stripes are enthusiastically encouraged to take up residence. But bring up the advantages of hunting wasps prowling about our yards, and there is generally a long pause. However, consider this: hunting wasps help in a very important task, the control of populations of plant-eating insects in gardens.

Hunting wasps include both vespids (social and solitary species) and sphecids (solitary). Along with a few other families of hunting wasps, the bees, and the ants, they are known as the Aculeata, or stinging insects, in the order Hymenoptera. They share the characteristic of chewing mouthparts but, apart from that, they are incredibly diverse in form and life history. In female hunting wasps, the ovipositor is often well-developed, and it may also function as a stinger. It is principally their diet that makes them potential garden allies.

California yellow jacket (*Vespula sulphurea*)

These wasps use those chewing mouthparts to prey on other insects, including many common garden pests.

Almost all wasps are carnivorous in the larval stage. Adult wasps prey on a variety of arthropods to provision their nests. They prey on the caterpillars of moths and butterflies as well as various flies, cockroaches, termites, and mantises. Even spiders may be fed to larvae. Adult wasps feed exclusively on nectar, pollen, and occasionally honeydew (the sugary waste product of insects such as aphids and scale insects). In a sunny, flower-filled garden, we often see adults visiting blooms of umbelliferous plants and flowers in the aster family. Gardens with a generous complement of native plants are sure to host many of the plant-eating insects that wasps hunt. Thus, the hunting wasps contribute to a healthy ecosystem.

The hunting wasps that provision their nests with prey for their larvae are broadly divided into two large groups. The solitary species may look formidable, but are generally not aggressive. These include mud daubers, thread-waisted wasps, and sand wasps. Social wasps such as the familiar hornets, paper wasps, and yellow jackets, can indeed be aggressive, most usually when protecting their nests.

Solitary species vary in size, from minute to some of our largest wasps. Most are ground nesters, though some nest in hollow plant stems, in existing holes in wood, or in mud nests that they build. Each female provisions her own nest, though they occasionally nest in aggregations. Some only prey on specific species, but many have a more general diet.

The behavior of solitary hunting wasps was elucidated in the late 1890s by French naturalist Jean-Henri Fabre. A keen observer, he studied insect behavior at a time when most entomologists worked on classifying insects, working with dead specimens. Fabre did not accept evolution, and tended to anthropomorphize liberally in his writings. However, his ten volumes of *The Life of Insects* captured the public imagination, and brought him high honors

Blue mud dauber (*Chalybion californicum*)

including a nomination for a Nobel prize. Many of his behavioral studies are still useful today. One of his most interesting discoveries was that female moths emit pheromones (chemicals released to evoke a reaction in an individual of the same species) to attract males.

Stinging wasps are undeniably a nuisance in some circumstances; generally the social vespid species are the culprits. They don't tend to attack us unless provoked, usually when we blunder into or harm their nest. "Pest" may be too strong of a word for many of these beneficial pest-eating insects. They have a lot in common with the social bees, which we don't revile or fear. Vespid colonies include queens, workers, and males (though unlike honey bees, these colonies only last for a single season). In common with most other hymenopterans, unfertilized wasp eggs develop into males, while only fertilized eggs develop into females.

Often, when we discover a wasp nest, our first instinct is to eliminate it. There are certainly plenty of toxic sprays for this purpose, but it is best to avoid

Western paper wasp (*Mischocyttarus flavitarsis*)

these due to the unintended collateral damage you may cause to other life in your garden. If an active nest in your yard is not in a high-traffic location, perhaps you can leave it be, so these avid hunters can help keep your garden's ecosystem in balance.

Admittedly, ground-nesting yellow jackets may pose more of a forbearance problem than those that are perched out of the way in trees or high on buildings. The common paper wasps, whose nests are often found under the eaves of houses, prey on caterpillars, a useful garden service. The bald-faced hornet, *Dolichovespula maculata*, builds its fabulous football-sized nest in trees or shrubs; adults of this species prey on yellow jackets and flies. Someone gave me one of these impressive nests (abandoned by its occupants) and it is wonderful for teaching. I love to watch people discover, as they examine the striated outer covering, that it is not humans but wasps that invented paper.

APPEARANCE Larvae are grublike or maggotlike. Females of aculeate wasps generally have a stinger. Many wasps have yellow and black aposematic (warning) color patterns. Size varies widely.

POPULAR NAMES Paper wasps, yellow jackets, hornets, thread-waisted wasps, mud daubers.

SCIENTIFIC NAMES Order Hymenoptera, families Vespidae and Sphecidae.

COMMON TAXA Two subfamilies of the Vespidae, the Vespinae and Polistinae, comprise all eusocial species, and include the familiar paper wasps, hornets, and yellow jackets. The thread-waisted wasps, the Sphecidae, are solitary. Most wasps in the Sphecidae have been moved to a different family, the Crabronidae, itself due to be again divided. Solitary subfamilies of vespids, the Eumeninae and the Masarinae, are discussed in the next essay.

DISTRIBUTION America north of Mexico, about 300 species Vespidae, 125 species Sphecidae, and 1225 species Crabronidae. Worldwide, about 5000 species Vespidae, 800 Sphecidae, and 9000 Crabronidae.

GARDEN ACTIVITIES Hunting wasps dine on caterpillars, flies, cockroaches, termites, mantises, and spiders, thus helping to control these populations in our yards.

Potters, Masons & Cuckoo Wasps

Some potter wasps suspend their egg from a slender thread at the opening of a beautifully constructed mud pot. Once hatched, the larva drops onto the paralyzed prey beneath—its first meal.

Much as wasps invented paper long before humankind came upon the utility of that now ubiquitous product, other wasps were working as potters and masons long before we discovered how to work clay. Many earthen wasp nests are small, and difficult to observe, hidden inside stems or far below ground. All of these insects have an annual life cycle, with the majority spending most of their time in their often distinctive nest in their immature stages.

Potter wasps (family Vespidae, subfamily Eumeninae) build their nests in the open as a single, graceful, narrow-necked earthen pot, a mere half-inch or less in diameter. The females form these lovely pots as housing for their larvae, on stems or sometimes on leaves or small twigs. Each pot is provisioned with a paralyzed caterpillar, spider, or beetle. A single egg is laid, positioned on the inside of the curved wall above the helpless prey, and then the top is carefully sealed. Marveling at the fragile and perfect miniature form of these vessels, it is easy to revert to a childhood belief in magical beings.

Potter wasps can be recognized by their curious double waist, an adaptation for bending their abdomen into the inside curve of their pots. Potter wasps'

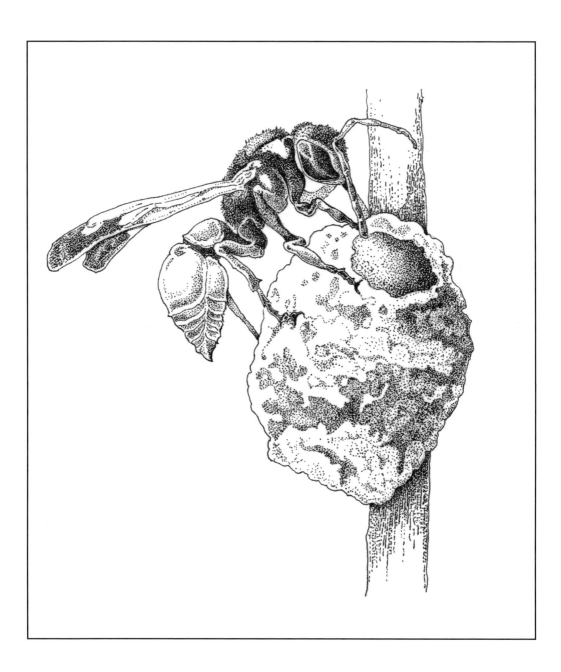

Potter wasp on nest (*Eumenes* sp.)

skill at capturing caterpillars is an asset in the habitat garden—if only we could teach them to feast exclusively on the cabbage butterfly larvae that attack our brassica crops!

Mason wasps are closely related to potter wasps. These use mud to build their homes, in abandoned beetle tunnels, nail holes, and other handy cavities. Both potters and masons are members of a branch of solitary wasps in the generally social vespids.

Spider wasps (family Pompilidae) are easily observed in the summer garden, sipping nectar from flowers (as do all the adult wasps of the groups discussed in this chapter). Usually dark-colored, often metallic with amber wings, a few are more brightly colored. They can be recognized by their behavior of flitting their wings as they run around on the ground seeking prey. Females have gracefully curled antennae. Solitary wasps, the larger members of this family hunt and paralyze correspondingly large spiders, which they drag into their underground nests.

The enormous tarantula hawks (*Pepsis* spp.) are our largest North American wasps; some species are over 4 inches in length! Although they are reputed to have a powerful sting, like other solitary wasps, they are generally unaggressive. The larvae of most pompilids feed on spiders. Adults build a nest in the ground, although some species paralyze a spider in its own burrow. Their usual technique is to stroke the spider's web to make it think it has caught prey, and then paralyze it with a sting before dragging it back to their nest to lay an egg on top of it. One day, when I pointed out to my young son the *Pepsis* wasps that were visiting the angelica blooms in our garden, he drew closer, his nose just inches from the flowers. How proud I was! The impressively large wasps went about their business of drinking nectar, and he got a good look at their remarkable metallic blue bodies with the amber wings—an experience he still remembers.

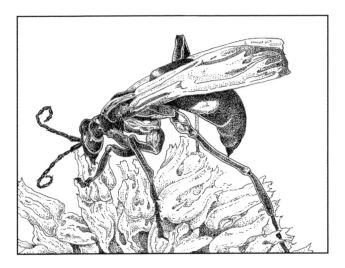

Tarantula hawk (*Pepsis thisbe*)

The cuckoo wasps (family Chrysididae) are generally small and often extraordinarily beautiful, bright shimmering metallic blue or green. The adults feed on flower nectar. They parasitize other insects, such as native bees and wasps, by laying their eggs in their victim's carefully prepared nest. Their larvae hatch and eat the food intended for the victim's offspring (this is the same phenomenon we encountered in bees, kleptoparasitism). The ovipositor does not function as a stinger, but cuckoo wasps have a concave abdomen and are able to defend themselves by rolling into a ball if threatened. While they resemble some metallic native bees, they lack pollen-gathering structures.

The so-called velvet ants (family Mutillidae) are infrequently spotted in gardens but always elicit surprise. Not ants at all, but wasps, the beautiful females, in shades of red, rust, gold, white, and black, run over the ground in arid areas of the West. Males, rarely seen, are winged. Both males and females are covered in a dense velvety coat. They pack a powerful sting, and some species

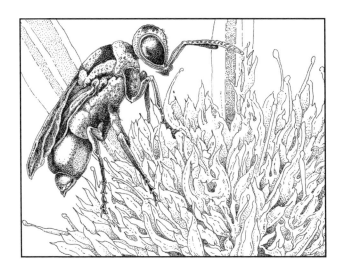

Cuckoo wasp (Chrysididae)

go by such daunting monikers as "cow killer." They are known to rub their legs or other body parts together to make squeaking sounds when disturbed (stridulate)—should you be foolish enough to disturb one. The mutilids are all parasites of insects, particularly of other wasps and bees.

The pollen wasps (family Vespidae, subfamily Masarinae) are an unusual group, limited to the arid areas of the western United States. They are unique among the wasps in provisioning their nests with pollen. These are black and yellow and, while some species closely resemble yellow jackets and other vespids, they are easily distinguished by their unique clubbed antennae. They build nests of mud or sand, attached to rocks or twigs.

One of the delights of exploring the garden is to seek out and find the various wasp nests, and match them with their occupants—but without getting too close! I encourage you to let these allies stay in your garden whenever possible, and let them provide some valuable pest-control services.

APPEARANCE Varies greatly within and between families. The most striking are the cuckoo or jewel wasps, which usually sport bright metallic colors. The pompilids are often dark-colored with very long rear legs. The Eumeninae often have typical black/yellow patterns. The Masarinae have clubbed antennae. The mutillids are generally covered in dense hairs, and females are wingless.

POPULAR NAMES Potters, masons, cuckoo or jewel wasps (ruby, emerald, gold wasps, etc.), spider wasps, velvet ants, pollen wasps.

SCIENTIFIC NAMES Family Vespidae, subfamilies Eumeninae and Masarinae. Families Mutillidae, Pompilidae, Chrysididae.

COMMON TAXA Potter and mason wasps, jewel wasp, tarantula hawks.

DISTRIBUTION In America north of Mexico, about 300 species pompilids, 227 species chrysidids (most diversity in arid areas of South and West), 225 species Eumeninae, 260 species Masarinae, and 435 species of Mutillidae (mostly in arid South and West).

GARDEN ACTIVITIES Hunting wasps, depending on their taxa, dine on a range of garden insects, including caterpillars, beetles, and more. Pompilids are spider hunters.

Helpful Hover Flies

One of the gardener's most useful allies—the larva of the diminutive chevron fly—was found in one study to exert from 70 to 100 percent control of aphids.

If you have ever paused near a patch of cosmos on a sunny summer afternoon, you are certainly familiar with the unique flight of the hover, flower, or syrphid fly (family Syrphidae). As it zigzags about the blossoms, occasionally stopping in mid-air, shimmering wings barely visible, there is no mistaking it. Its ability to hang suspended in mid-air explains my preferred moniker, hover fly.

Unlike other winged insects, flies have only one pair of flying wings, while the second pair is reduced to two little knobs, the "halteres." These function like miniscule gyroscopes, allowing flies to quickly change direction, a useful ability when being chased by predators, or a biped armed with a swatter. Not that the grateful gardener should want to swat most hover flies! With very few exceptions, these flies are among our most abundant and best garden allies, both as pollinators and as predators of pest insects.

Even for those familiar with hover flies, a closer look will reveal a surprising diversity of species. Most are in the ¼- to ½-inch range. Many mimic bees, as that coloring advertises danger. You can quickly distinguish them from bees, however, by their shorter antennae, stout waist, and single pair of wings (bees have two pairs). Common species may be slender and cylindrical, or have a wide, flattened abdomen. While a few are fuzzy, most have glossy bodies. A defining characteristic of this family is what is called a "spurious" wing vein, a fold in the middle of the front wing membrane, best seen with a hand lens.

Allograpta obliqua

When the larvae hatch, they are easily mistaken for pests—so look closely. They resemble small slugs or caterpillars, often smooth and light green with a longitudinal white stripe. Some species are brown, yellow, spiny, and even transparent. Having no eyes, flower fly larvae were long thought to blindly detect their prey by raising the front half of their body and swinging it from side to side until they bumped into a tasty morsel. More recent evidence shows that at least some species detect their prey by sensing chemicals.

The appetite of these larvae can be prodigious. Aphids are a principal food source of the majority of common garden hover flies. A few other species prey on leaf beetle larvae, eggs, or other insects. Some have been estimated to eat up to 1200 aphids in the larval stage! Once its victim is captured, the larva holds its prey in the air, sucking it dry of juices, while nearby aphids or beetle larvae appear to pay no attention to the demise of their unfortunate neighbor.

Not all hover flies prey on pests. Some species' larvae dwell in the nests of social insects such as bees, wasps, and ants, where they live as scavengers. A few eat

Paragus tibialis

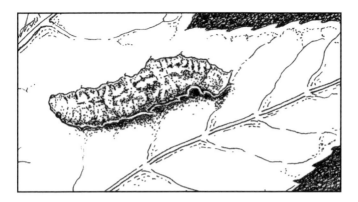

Larva of *Allograpta obliqua*

roots, stems, and bulbs. But for the most part, these exceptions are of no concern to us, as gardens are unlikely to suffer any significant damage. Those that frequent decaying organic matter aid in decomposition.

Much of what we know about hover flies is properly credited to the British, who have a particularly well-studied insect fauna. I received satisfying confirmation that they thrive in complex settings in the pages of *The Ecology of a Garden* by Jennifer Owen. Simply put, offering them a diversity of flowering plants (found in most gardens anyway) is the ideal way to attract them.

Owens also observed that hover flies are vulnerable to wind (as are many flying insects); therefore if you garden on a windy site, provide some shelter such as fencing or hedges. Predatory species seek out pest colonies in which to lay their eggs. Females of aphid-eating species are bound to find a multitude of egg-laying sites in any garden. As for their life spans, some produce but a single generation annually (univoltine), while others have several generations annually (multivoltine); this varies with geographic location. Many have a long flight season, and some overwinter as adults. Many pupate and overwinter in leaf litter, yet another reason to not rake away all the autumn leaves in your garden.

Adult hover flies feed on nectar and pollen and occasionally on the honeydew excreted by aphids. They need nectar to fuel their high-energy flight and females need pollen to produce eggs. Planning a long season of bloom is the best way to attract them. To control early outbreaks of aphids, you'll want to include plants that provide early-spring pollen, such as willows, grasses, sedges, and wildflowers. Include small-flowered plants such as alyssum and the more diminutive members of the aster family, as the smaller blooms attract many smaller species. Perhaps the most effective garden flowers for late season are the beautiful fall asters. The helpful hover flies love a garden full of flowers, and so do I!

APPEARANCE Larvae resemble tiny caterpillars or slugs. Adults may be quite small, perhaps 0.1–0.14 in. (3–4 mm), up to 1.25 in. (20 mm). Variable colors, often marked with yellow or white bars or spots on a dark background, short antennae, stout waists.

POPULAR NAMES Flower flies, hover flies, syrphid flies.

SCIENTIFIC NAMES Order Diptera (flies), family Syrphidae; divided into four subfamilies, of which only the Syrphini and Pipizini eat aphids.

COMMON TAXA Aphid-eating species include: chevroned hover fly (*Allograpta obliqua*), large hover fly (*Scaeva affinis*), and *Sphaerophoria*, *Paragus*, and *Toxomerus* species.

DISTRIBUTION 813 species in America north of Mexico.

GARDEN ACTIVITIES Primarily prey on herbivorous insects. Larvae of some almost exclusively aphid-eaters, but a few eat other insects, including leaf beetle larvae, whiteflies, and moth caterpillars. Adults feed on nectar and pollen; a few are attracted to honeydew exuded by aphids and others. The narcissus fly feeds on daffodil bulbs and can be a pest under some circumstances. Those feeding on decomposing plant matter and plants are in the subfamily Eristalinae.

Tachinid Flies

Those bristly flies feeding on the nectar of your garden flowers aren't house flies, but some of our best garden allies! Tachinid flies parasitize many arthropods, mostly insects. Hosts include caterpillars, sawflies, orthopterans, mantises, cockroaches, flies, earwigs, and true bugs.

My husband once presented me with an enormous tachinid fly that he had caught at work. What an inspired gift! I was far more impressed with the fact that he snagged such a robust and rapid flyer with a small-mouthed bottle than I would have been with, say, an easily purchased bouquet.

The bristly fly he caught, *Hystricia abrupta*, is the largest West Coast parasitoid fly and readily recognizable (an entomology student commented that it looked like it should be riding a Harley, and I have ever after called it "Harry Tachina"). Most tachinids, however, are not so easily discerned from other flies. Many superficially resemble a common house fly, although more bristly. The characteristic bristles are sensory organs, sensitive to air movement and used in orientation and navigation.

All tachinids are parasitoids, that is, they deposit their eggs on or near host arthropods, and the larvae enter and consume the host insect as they both develop. Parasitoids almost invariably kill their host. Because most do not have a piercing ovipositor, they have evolved other strategies for egg-laying. Larval development is typically from one to three weeks, but some species have a prolonged larval stage over many months. Tachinids emerge from

Hystricia abrupta

Trichopoda pennipes

their host to pupate in soil or leaf litter. As adults, many of the common garden tachinids are flower visitors, feeding on nectar and pollen. They tend to be
diurnal or crepuscular, that is, active at dawn and dusk.

Tachinids are a perfect example of why it is a good idea to develop overall
insect diversity in our gardens. Most are more or less generalists. In their larval
stages they parasitize a broad range of insect hosts, among them beetle larvae
(especially scarabs and leaf beetles), various bugs, grasshoppers and crickets, and
caterpillars of butterflies and moths. Many specialize on sawflies, common and
vexing leaf-eaters that damage some of our favorite shrubs and trees. They are

Voria ruralis

also known to attack centipedes. Tachinids are receiving increasing attention from researchers who study biological control, as they appear to be equivalent in effectiveness to the more commonly used parasitic hymenopterans.

A few tachinids are regarded as pest species. For example, the uzi fly (*Exorista bombycis*) is the bane of silkworm ranchers. For those who like to observe the life cycle of monarch butterflies, few sights are as disheartening as seeing a plump maggot drop out of a caterpillar on a silken line, leaving behind a shriveled exoskeleton.

Some tachinids have evolved complex and fascinating strategies for locating their hosts. For instance, the feather-legged fly (*Trichopoda pennipes*) locates its preferred prey, primarily squash and stink bugs, through sensing the pheromone produced by their male prey to lure females of the species. *Ormia ochracea*, a small nocturnal fly, uses the mating call of a male cricket to track down its host; the cricket, while attracting its mate, is simultaneously attracting its doom. *Ormia* is notable for its directional hearing—second to none among animals—a result of its uniquely coupled ears. Interestingly, researchers have used this remarkable trait to develop low-noise differential microphones, to the great benefit of the hearing-impaired community.

Tachinids exhibit between them a great breadth of host range, in 11 orders of insects plus other arthropods. This variation in host range makes them useful in studies of evolutionary biology and ecology, and advance our understanding of basic biological processes. I think of this, as I pause in my garden to watch a *Trichopoda pennipes*, that small black- and amber-colored fly possessing elegant polished black legs with large "combs" on its rear femurs, bright against the verdant leaves of a giant polygonum. I am glad that they are being studied, but in my garden it is enough to include a diversity of plant species to attract them. Tachinids have a welcome home in my landscape, where I am grateful for their sophisticated population-regulating adaptations, leaving me free to simply enjoy their frequent visits.

APPEARANCE These often resemble houseflies, though many are smaller. Adult size varies, 0.07–0.75+ in. (2–20 mm). Many species are black or gray; some have vivid markings in yellow, orange, red, and even metallic blue or green. Some are wasp and bee mimics. Eggs are small white ovals, laid singly. Larvae are maggots (legless larva).

POPULAR NAMES Tachina flies, tachinids.

SCIENTIFIC NAMES Order Diptera, family Tachinidae, part of the superfamily Oestroidea, which includes flesh flies (Sarcophagidae), blow flies (Calliphoridae), and bot flies (Oestridae).

COMMON TAXA Feather-legged fly (*Trichopoda pennipes*), *Voria ruralis*, *Erynnia tortricis*.

DISTRIBUTION Second largest dipteran family. About 1350 species; many remain to be described.

GARDEN ACTIVITIES They attack many injurious insects (only parasitoid wasps have a greater impact).

3

DIGGING DEEPER

Predators and Parasites

Insect predators and parasites are legion, and undoubtedly active in your garden right now, attacking and killing unwanted visitors, and you rarely know it. It's a "bug-eat-bug" world. Remember, for every insect that has ever plagued your flowers, vegetables, or fruit, there exists at least one natural enemy, usually several.

Predation is straightforward. A predator kills and eats many individuals over the course of its life. Also it is usually bigger, not smaller, than its intended prey. Avoiding predation is a strong force in nature, specifically in natural selection. Many insects have evolved defenses against predators. For example, velvet "ants," actually hunting wasps, touched on in the previous chapter, squeak using a "stridulatory" organ, have a painful sting, display a warning coloring (a characteristic known as aposematism), have an alarm secretion, *and* have a very hard exoskeleton—not exactly the most appetizing prospect. The predators, in turn, respond to the new selection pressure, and we're off to the races! Those velvet ants have won their particular race. It is a rarity, but *no* predators are willing to tangle with them.

The biology of parasitism is more complex. Maria Sibylla Merian, an early naturalist and artist (1647–1717), observed her subjects closely, particularly caterpillars and the plants on which they feed. Well ahead of her time, she was one of the first to record the existence of parasitoids. Many of her paintings beautifully illustrate the complexity of the food web, depicting not only a plant, but also the associated insects in all stages of development. More prosaically (or graphically!), insect predation and parasitism has been inspirational in science fiction—witness for example Ridley Scott's classic 1979 film *Alien*.

Before we go any further, let's clarify the difference between a parasite and a parasitoid. A parasite lives on or inside a host, and is usually far smaller. Although it may move from host to host, it very rarely kills its comfortable abode. Indeed, parasites are often bound to their host for life. In contrast, a parasitoid may be much larger than a parasite. A parasitoid also lives on or inside a host, but only for part of its life cycle, with a free-living adult stage. Those that develop within their insect host are endoparasitoids; ectoparasitoids develop externally to their host. Parasitoids almost invariably kill their host.

Some insects simply act as parasites on other animals. For example, the infamous bot flies, which parasitize mammals in the flesh or digestive tract, cause discomfort at best and infection and even death at worst. Parasitism in insects themselves is often overlooked. This is partly because the hosts may appear healthy to the casual observer, and partly because the creature doing the damage is so small or out of sight. One parasite we do notice is the varroa mites that plague honey bee colonies. Some parasitoids pupate externally, but many drop to the ground. Others complete their life cycle inside the host's body, emerging only once they reach their adult stage. In this chapter, we will examine flies and wasps that operate in these ways in our gardens.

If all you know of flies is chasing house flies with a swatter, it is time to take a closer look at the flies in your garden. There is more variety in the order Diptera than most people imagine. Among them are a number of predaceous and parasitic species. We were introduced to the hover flies (predators) and the tachinid flies (endoparasitoids) in the previous chapter on flower visitors.

In the amazingly diverse and large insect order Hymenoptera (which includes wasps, bees, and ants), it is the wasps that play a principal role as parasitoids. When most people hear the word "wasp," some sort of pesky insect comes to mind, such as the yellow jackets that harass us at a picnic or the paper wasps that dangle a threateningly large nest over a doorway. Despite the fact that my dad was a parasitologist (he specialized in tropical diseases), it was only later, in a college entomology class, that I was introduced to the Parasitica, the parasitic Hymenoptera, so rich in wonderful adaptations. With over 89,000 species worldwide (over 60 percent of all wasp species), I was hooked! The most common parasitoid wasps in gardens include the braconids, the ichneumonids, and several families of chalcids. It is a powerful trio, essential to our ecosystems.

When we delve deep enough, concepts of predation and parasitism begin to overlap. This is most easily puzzled over in the Hymenoptera, where distinctions are not always clear. The stinging wasps, for instance, are predatory hunters, capturing and moving their paralyzed prey to a nest. As hunters, they are not always wedded to only one kind of prey. They often lay their eggs in the vicinity their victim, but never in its body. So are the larvae functioning as predators or parasitoids? Hmmm. Nature, I may have mentioned, is complicated!

Predators and More

Garden allies can have curious, and even gross (to us) life cycles. Some predatory midges exhibit paedogenesis, the female reproducing while she is still in larval form. In the case of the predatory midges, the larvae she carries internally eventually eat their way out of her body.

Flies, as you've likely noticed, are all around us. The hover flies (predators) and the bristly tachinids (parasitoids) are familiar sights to most observant gardeners, but many other interesting flies frequent our yards. Over 40 families of flies function as predators of other insects. Some predaceous flies feed on dead organic material (detritivores) or waste (coprophages), activities that are a benefit to our gardens.

I know the weather is warming when I first spot the brilliant emerald sheen of the long-legged fly (family Dolichopodidae), usually perched on a leaf in my yard. It isn't actually the male fly's beautiful metallic colors of green, blue, bronze, silver, or gold that attract the female's attention, but instead—you guessed it—his long, graceful legs! But in case that is not enough to lure a female, many long-legged flies also provide "nuptial gifts," usually prey, which they dangle in front of the desired female. Predaceous in both the larval and adult stages, their prey includes unwelcome plant-damaging insects such as thrips, aphids, various small larvae, and spider mites. The closely related predaceous dance flies perform a synchronized dance in their mating swarms, and also exhibit the trait of offering nuptial gifts.

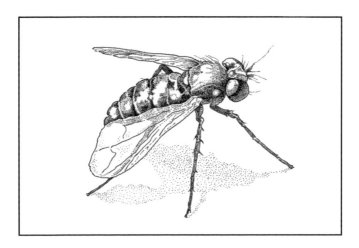

Long-legged fly (*Condylostylus longicornis*)

To the casual observer, the predatory midges (family Cecidomyidae) resemble tiny mosquitoes, albeit with their antennae curled back over their heads. They are in fact tiny flies (as are mosquitoes). While a few eat plants or cause galls, most are active allies. Their larvae are specialist or general predators (a few are parasitoids) of soft-bodied insects, especially scale insects and members of the mealybug family. Others prey on whiteflies and thrips. The tangerine-colored larvae of aphid flies (*Aphidoletes*) look like miniature slugs. They eat only aphids and lots of them, and are such reliable specialists that they are used for biocontrol of aphids in commercial greenhouses. In your garden, chances are if you have aphids, predatory midges are already present. Look for their characteristic orange larvae on the undersides of leaves of afflicted plants.

One hot summer day, driving through the desert, one of the biggest robber flies (family Asilidae) I have ever seen flew into our open car window, much to everyone's consternation. It is true that some robber flies can reach formidable size, and ambush (hence their name) a wide range of large prey such as

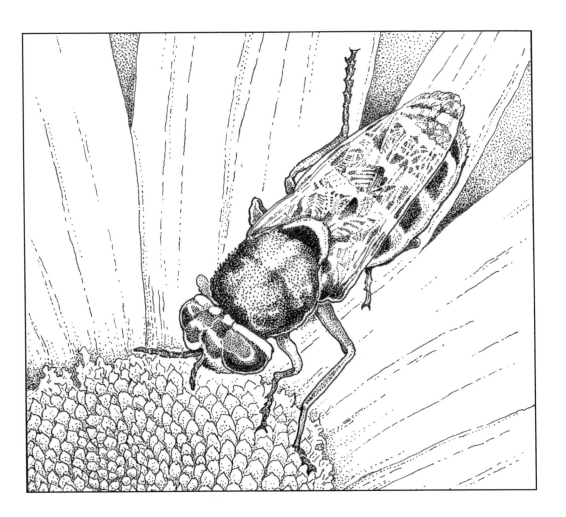

Soldier fly (*Odontomyia* sp.)

grasshoppers, stinging wasps, dragonflies, and spiders. Smaller ones attack correspondingly smaller prey, but the fearsome robber flies can catch prey larger than themselves, sometimes far larger. They are powerful, quick in flight, and able to catch prey on the wing, sallying forth with remarkable speed from a convenient perch. Prey is quickly subdued with a venomous bite. Robber flies can be recognized by their rapid flight, a generally tapered abdomen, and especially by their bristly "mustache" (the mystax), thought to protect them from flailing prey. Many are wasp mimics. These large flies, whether pinballing around the confines of your car or zipping around your garden, are no threat to humans. They can bite, but first you would have to catch one!

The stiletto flies (family Therevidae) are closely related and similar to robber flies, although these are usually smaller and more slender. Unlike the robber flies, they are clean-shaven. In their larval stage, they prey on soil arthropods. Adults dine on nectar, pollen, and honeydew.

You may also encounter soldier flies (family Stratiomyidae) from time to time. Many are wasp mimics. The most commonly encountered species is the shiny black soldier fly (*Hermetia illucens*). A sphecid wasp mimic, it has two transparent "windows" on its abdomen that give the appearance of a slender waist. Adults are usually nectar feeders. Eggs are laid in batches of hundreds, and dark grayish-brown larvae soon hatch. These mainly break down decomposing organic matter in the soil. They also prey on soil arthropods, especially the larvae of blow flies and house flies. Voracious eaters of soggy fruit, vegetable waste, and coffee grounds, soldier fly larvae are also helpful in home composting systems (while they will show up on their own, they can be purchased live and given their own compost bin). Laden with fat and protein, the larvae are used in aquaculture and for pet and human food. My chickens love dried soldier fly larvae!

As you might expect, the golden dung fly (family Scathophagidae) is found on dung, in this case usually that of large mammals, including many of our domesticated livestock. They are hairy flies, bright golden yellow with furry front legs. The males, unlike most insects, are generally larger than the females. Adult dung flies prey on small insects, while the larvae eat animal waste (coprophages). Females forage in vegetation, only visiting dung to mate and lay eggs. The males, on the other hand, spend most of their time on dung, waiting for females to make an appearance.

There is still much to be studied in the large and fascinating realm of flies. I reach for one of the treasures on my bookshelf, *The Secret Life of Flies* by Erica McAlister, to delve into the mysteries and wonders of this most highly evolved of the insects, with all its curious adaptations. My family would prefer to not study large desert species from the inside of a vacation-loaded car, but is happy to enjoy watching them with me from the comfort of our garden chairs.

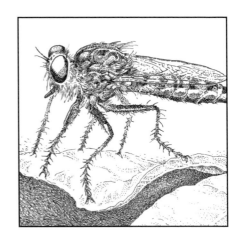

Robber fly (*Asilidae* sp.)

APPEARANCE Single pair of wings (unless wingless); second pair has been reduced to halteres. They range in size from a mere 0.03 in. (1 mm; Dolichopodidae and Empididae) to almost 2 in. (50 mm; the largest Asilidae).

POPULAR NAMES Long-legged and dance flies, midges, robber flies, stiletto flies, soldier flies, and golden dung flies.

SCIENTIFIC NAMES Diptera includes about 170,000 described species in about 110 families; there may be more than one million species worldwide.

COMMON TAXA Families with predatory members discussed here include: Dolichopodidae, Empididae, Cecidomyiidae, Asilidae, Therevidae, Stratiomyidae, Scathophagidae. Many other fly families include predators.

DISTRIBUTION Species widely found in America north of Mexico: 1300 Dolichopodidae, 460 Empididae, 1000 Cecidomyiidae, 1040 Asilidae, 150 Therevidae, 250 Stratiomyidae, 150 Scathophagidae.

GARDEN ACTIVITIES Predatory flies dine on a great variety of insect pests. As adults, many flies visit flowers, feeding on nectar and sometimes pollen.

Nosing Around with the Parasitoids

The spider flies are parasitoids of spiders. Some have a "tongue" (proboscis) so long that when the fly folds it back underneath its abdomen, it juts out the other end, looking like a respectable stinger!

Having looked at the hunters, let us now consider some parasitoids in the world of flies. At least some are bound to be present in your garden. Some are highly specialized, and can even be restricted to a single species of host. Others are not as discriminating. Females lay eggs inside or on a host, and the hatched larvae feed and grow on the living host. Their presence is a mixed bag for gardeners. Some of their prey are desirable garden insects such as some of our most beautiful butterflies, but they also parasitize a great many unwanted insects.

Tachinid fly larvae develop within their insect host (they are endoparasitoids). They do not, however, prevent the host from continuing its own growth and development. The term for this is "koinobionts," a phenomenon more often observed in the parasitoid wasps than flies. The victim continues to grow and eat, even when parasitized in an early larval stage. (In contrast, an idiobiont is a parasitoid that halts the growth and development of its host, usually by injecting a paralyzing toxin.) The parasitoid usually emerges from the pupal stage of the victim, by which time the hapless host has died. This means that

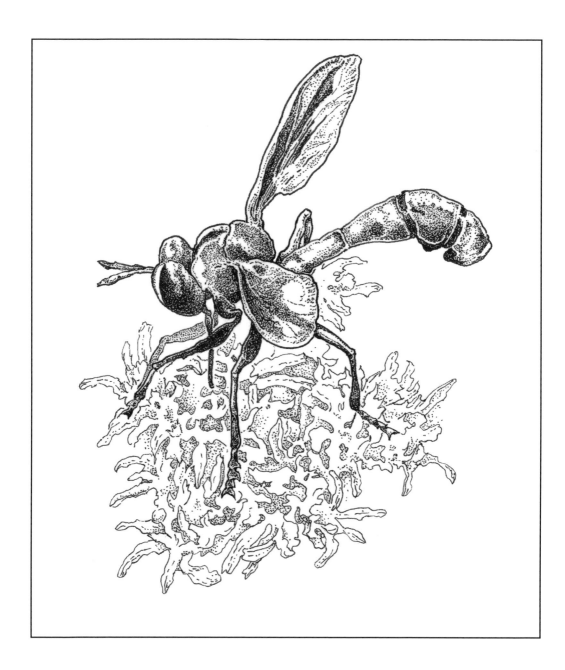

Thick-headed fly (*Physocephala burgessi*)

host populations are not always controlled by tachinids in the first generation, although subsequent generations may be suppressed.

Most tachinids, as mentioned earlier, have a broad host range, parasitizing all sorts of insects. This is unusual for parasitoids and certainly boosts their value as garden allies. *Compsilura concinnata*, for instance, attacks almost 200 species of insects—including the dreaded gypsy moth. While some tachinids lay eggs directly on their host, others lay thousands of micro-eggs on leaves, where they are ingested by the host. Some incubate their eggs internally, and lay eggs with fully formed larvae that emerge immediately when laid on or near their hosts. Others have larvae that go through progressive stages of development and their first stage (first "instar"), called a planidia, is actually mobile and able to hunt for a host right after hatching—first things first! Later instars are sedentary, and adapted for eating.

An interesting member of the food web is a taxon of odd-looking little insects called thick-headed flies (family Conopidae), named for their broad head. Many are mimics, resembling wasps or bees, causing them to be noticed more often than some common flies. Nectar feeders as adults, in the larval stage most are endoparasitoids. They intercept their intended host in flight, and pry open the segments of the abdomen to deposit an egg with a special structure called a theca, commonly and graphically referred to as a "can opener." One subfamily has barbed eggs it shoots onto the host with a harpoonlike ovipositor! They parasitize a broad range of insects, such as honey bees, bumble bees, social wasps, cockroaches, and some flies. Look for adults on many common garden flowers.

The big-headed flies (family Pipunculidae) are easily recognized by the huge eyes that make up most of their head, providing correspondingly good vision, and by their typically black coloration. They are mostly parasitoids on hoppers, spittlebugs, cicadas, and a few on crane flies. Females inject a single egg into nymphal or adult hoppers (or into crane flies). The larva feeds

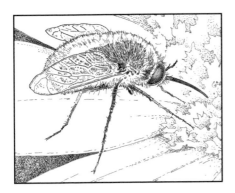

Bee fly (Bombyliidae)

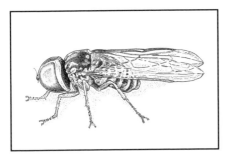

Big-headed fly (Pipunculidae)

internally and exits to pupate; the life cycle is completed in soil or leaf litter. The adults feed on honeydew secretions. You may spot them hovering, much like hover flies, over vegetation in your garden.

With their furry round bodies, it is easy to mistake the aptly named bee flies (family Bombyliidae) for bees. Fairly common in gardens, adults hover above flowers, prodding for nectar with their long straight proboscis. When at rest, their wings are held at a backwards angle characteristic of flies. They are sometimes important pollinators, especially on flowers with long narrow floral tubes such as verbena, lungwort, flowers in the mustard family, lavender, thistles, and many more.

Bee flies are ectoparasitoids, developing externally on a host's body. Like some of the tachinids, they are hypermetamorphic—that is, the initial mobile larval stage (planidia) is very different in appearance from the larval stages that follow and feed on the host. Although they parasitize a variety of insects and spiders, their preferred hosts are bees and wasps. They usually deposit eggs in the vicinity of their hosts' burrows or nests. But sometimes they literally fling their eggs at the entrance, a behavior that has earned them the moniker "bomber flies." I have seen many interesting dramas while puttering in my garden, but I hold out hope that I may one day witness a bomber fly in action!

APPEARANCE Many flies have bristles; tachinids are especially bristly. Predatory flies are often bee and wasp mimics. Tachinidae, 0.08–0.7 in. (2–18 mm), Conopidae, 0.16–0.7 in. (4–18 mm), Pipunculidae, 0.06–0.33 in. (2–10 mm), and Bombyliidae, 0.08–1.6 in. (2–40 mm).

POPULAR NAMES Tachinid flies, bee flies, thick-headed flies, big-headed flies.

SCIENTIFIC NAMES The order Diptera includes dozens of families. Worldwide, there may be over one million fly species; only about 125,000 have been described and named.

COMMON TAXA Families with parasitoid members discussed here: Tachinidae, Conopidae, Pipunculidae, and Bombyliidae. Many other fly families include parasitoids; *Cryptochetum iceryae*, which attacks cottony cushion scale, is in the Cryptochetidae.

DISTRIBUTION In America north of Mexico, species of these families are commonly found: 1300 Tachinidae, 70 Conopidae, 130 Pipunculidae, and 800 Bombyliidae.

GARDEN ACTIVITIES Parasitoid flies attack a great variety of arthropods, mostly insects; many are specialists. Some parasitoid adult flies visit flowers, feeding on nectar and occasionally pollen.

Small but Mighty

Look closely at a colony of aphids. Do you see hard brown motionless aphids? They are called "aphid mummies," and have been parasitized by minute braconid wasps.

A great many wasps are innocuous garden residents, usually nonaggressive, and in many cases, stingless. Most parasitoid wasps are smaller, often far smaller, than their more well-known relatives, the stinging (aculeate) wasps. Many aphid parasitoid wasps, for instance, are so tiny that they go about their business virtually unnoticed, but their presence is easily ascertained by examining an aphid colony. Look for papery black or beige "mummies," many with a neat exit hole gnawed by an emerging adult wasp.

The non-stinging wasps in the large and diverse braconid family parasitize caterpillars, flies, true bugs, and beetles, in addition to aphids. They are quite small, but at a quarter-inch to a half-inch, some species can be spotted in the garden. You may recognize braconids as the slender dark wasps commonly found hovering around aphid colonies or caterpillars. (In general, virtually all tiny wasps found lingering in the vicinity of herbivorous insects, regardless of identity, are garden allies.)

There are many specialists. Braconid aphid parasitoids pupate inside their luckless host and emerge as adults. The specialist *Diaeretiella rapae* is a welcome and common garden resident, reducing populations of the cabbage aphids that attack some of our favorite vegetables, not just cabbage but also other

Hornworm with wasp pupae (*Apanteles congregata*)

Apanteles sp.

Diaeretiella rapae ovipositing on a cabbage aphid

mustard-family members such as broccoli, collards, and arugula. Other species of braconids specialize on aphids that are uniquely found on beans (legumes), roses, and other garden plants.

But many braconids, especially those parasitizing caterpillars, emerge to pupate outside the host. Caterpillars may host numerous individuals (these parasitoids are termed "gregarious"), so should you come across a caterpillar in your garden festooned with small silky cocoons, you will know braconids are at work. The reviled tomato hornworm is attacked by *Apanteles congregata*. The imported cabbageworm is parasitized by *Apanteles glomeratus*. When you're out doing the messy business of picking off the unwanted caterpillars and worms from valued plants, you can spare these afflicted individuals. They're goners. The parasites will complete their life cycle and kill their host, increasing the population for the next season.

Depending on the species, braconid wasps may be egg, larval, or pupal parasitoids; all phases can be helpful in controlling pest populations in our gardens. Those that attack the eggs of their prey are devastating, and are the best allies. These deposit eggs on (or in) the egg stage of the host. Their larvae develop inside the egg, usually thwarting the host's ability to hatch. In some species of hosts, the egg does hatch, but the victim has little opportunity to damage any plants before its untimely end.

Many braconids attack the larval stage of their host. Although some caterpillars quickly lose their appetite for your prized garden plants once parasitized, I am sorry to report that others continue to eat, apparently oblivious to their internal guest. Some will even increase their plant-damaging activity when parasitized! The familiar imported cabbageworm, when afflicted, has been found to eat up to one and a half times the usual amount of food over its larval lifespan. Still, most larval-stage parasitoids offer a more immediate benefit, and in addition parasitized larvae will not survive to reproduce.

Pupal parasitoids, those that attack the pupal stage of their hosts, are a different story. The benefit, in such a case, is delayed until the next generation, as the host larva has already done its damage. Yes, this may seem like meager comfort as you watch your kale being munched. But be patient, for relief lies in waiting for populations of these helpful little insects to build up in your garden. This is why, counterintuitive though it may sound, you do not want to kill off and dispose of obviously parasitized pest caterpillars. These can often be recognized by the blackened spot where a braconid laid its egg.

Many of these parasitoid wasps are nectar feeders as adults—news a gardener can use. Capitalize on that and welcome these garden allies to your yard by planting their favored plants. They visit umbellifers (family Apiaceae), with umbrellalike heads of tiny flowers; good candidates include dill, parsley, cilantro, fennel, angelica, white lace flower, blue sea holly, and more, plus other small-flowered plants such as yarrow and alyssum. The flowers also provide a banquet for lady beetles, lacewings, hover flies, soldier beetles, native bees, and legions of other garden allies. In the case of the umbelliferous herbs I mention, you might have been in the habit of harvesting their foliage for kitchen use and then removing the plants. I recommend leaving a few to bloom. A garden full of useful and ornamental umbellifers is sure to please both beneficial insects and gardener alike!

APPEARANCE External cocoons on hosts are usually white or yellowish, with a "fluffy" appearance; the most easily spotted sign of braconids. Rarely over 0.6 in. (15 mm), the smallest are 0.04 in. (1 mm). Slender or stout bodies; long antennae. Wings often have a dark spot on the leading edge of the forewing. Color ranges black through brown, reddish, and yellowish tones.

POPULAR NAMES Braconid wasps.

SCIENTIFIC NAMES Order Hymenoptera, family Braconidae.

COMMON TAXA The majority are native insects that parasitize Lepidoptera. *Aphidius colemani* and *A. ervi* parasitize aphids. Some widely found beneficial species are introductions.

DISTRIBUTION Over 1700 known species in America north of Mexico; may be over 100,000 worldwide. Many species remain undescribed.

GARDEN ACTIVITIES Braconids parasitize a wide range of hosts, including many of our worst garden pests such as aphids, leaf miners, cabbage loopers; other victims are beetle larvae, flies, and sawflies. They often parasitize lepidopteran larvae such as cabbage butterflies and gypsy moths. As adults, these wasps often seek nectar.

ICHNEUMONID WASPS

Not Icky at All!

The formidable appearance of the western giant ichneumon wasp, with its long ovipositor, is only a threat to its host, as it is unable to sting. Instead, the dangerous-looking appendage drills into trees to deposit eggs. When they hatch, the larvae attack wood wasps and beetle larvae.

Ichneumonid ("ick-new-mon-id") wasps are among the most common winged insects found in gardens. Few know their names, although some common species are readily recognized. Slender, thread-waisted, and often with abdomens that are flattened from side to side, these parasitoids range in size, but most are quite small, less than ⅛ inch long (though some can be up to 5 inches). They come in myriad colors. Long antennae, along with a slender waist and transparent wings, are the distinguishing characteristics. The very smallest have brief lifespans, sometimes only a week or two, but may have several generations per year. Other species have just one generation annually.

They parasitize an enormous number of herbivorous insects, attacking their prey in larval and pupal stages and usually developing inside the host (endoparasitoids). Almost always, a single larva parasitizes a single host, although in some cases there will be multiple larvae in a single host. The larva of the useful wasp named *Hyposoter exiguae*, for example, spins its single cocoon in several species of crop pest caterpillars, including armyworms, cabbage looper, and tussock moth. Other ichneumonid species lay multiple eggs, resulting in multiple larvae in a single host ("gregarious parasitoids").

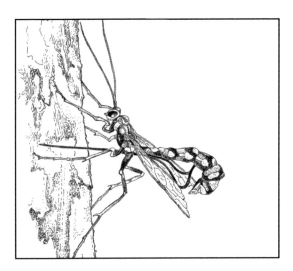

Western giant ichneumon (*Megarhyssa nortoni*)

Ichneumonids can have a significant impact on many garden insects. Among their favored targets are flies, grubs of wood-boring beetles and weevils, leaf miners, aphids, and other winged insects, including other ichneumonid wasps. These wasps may be habitat specialists, for example exclusively seeking out leaf rollers or wood-boring insects, regardless of species. Many useful species attack sawflies and other agricultural pests. Full disclosure: they also attack many beautiful and desirable butterfly and moth species, in addition to some valued spiders' egg cases. These are the complexities of a healthy garden, and why I try to avoid thinking in terms of "pest" and "beneficial."

Interestingly, these parasitoids are sometimes themselves parasitized! When this happens, we call them hyperparasitoids, a term first coined by Jean-Henri Fabre, the early French naturalist who closely observed insect behavior and discovered this extraordinary biological relationship. Secondary parasitoids can

reduce the beneficial effect of a primary parasitoid, but in our gardens, as in the wild, it is all just part of the web of life.

Quite a few ichneumonids are only seen in wild habitats, but gardens still get their share. *Ophion* species are common and familiar even to non-gardeners, as the adults are often attracted to lights in the evening hours. In the garden, they are seen walking about on plants with a characteristic staccato movement, their long antennae in constant motion. What are they doing? They are searching for suitable host insects in which to lay an egg. *Ophion* species parasitize scarab grubs such as June beetles, and the caterpillars of numerous species of moths.

For the most part, these wasps are harmless and nonaggressive. Although many have formidable-looking ovipositors, they are unable to sting; these are used to probe for wood-boring larvae. The majority are day flyers. Those with short ovipositors may be able to "sting" if handled, but have no venom to inject.

Some are active early in the morning, others are out and about later in the day. Look for adults visiting flowers for nectar.

Frequently they are habitat specialists, as opposed to prey specialists—this is something we gardeners can capitalize upon. That is to say, they are very particular about the niche in which they search for hosts, rather than specializing on particular prey. Encouraging high plant and habitat diversity will create niches for them.

APPEARANCE Size varies, 0.1–1.5 in. (3–40 mm). Pupae are usually external, smooth, oval cocoons. Adult ichneumonids range in size; some are impressively large. They are recognized by their slender waists, long antennae, and transparent wings. They are differentiated from the closely related braconids by wing venation (not easily observed without collection). Color ranges from black through brown, reddish, and yellowish tones. Many have long ovipositors, often longer than the body.

POPULAR NAMES Ichneumonid wasps, ichneumonid flies.

SCIENTIFIC NAMES Order Hymenoptera, family Ichneumonidae.

COMMON TAXA Many species may be found in a single garden. Common orange-colored *Ophion* species are often seen at lights.

DISTRIBUTION Over 5000 in Nearctic Region, possibly 3000 more. Arguably the largest animal family; definitely the most species-rich hymenopteran family.

GARDEN ACTIVITIES Most specialize on caterpillars (Lepidoptera larvae); others parasitize aphids, flies, beetles, or spiders.

It's All Relative: A Superfamily

The males of some species of Trichogrammatidae wasps are wingless. They mate with their sisters inside the host egg, and then die without ever leaving the confines of their host.

When I was conducting my graduate research, I visited Bob Zuparko, an expert on parasitic wasps at the California Academy of Sciences. I proudly put forth my collection of minute wasps. After examining them, he said "let me show you some *really* tiny parasitic wasps." He pulled out a slide and slipped it under a microscope. Imagine my surprise to see upwards of 50 wasps under a single cover slip. I realized then that some of what I had thought were "dust motes" floating around my garden are in fact insects! These miniscule creatures seem to push the boundaries of how small an animal can be and still have digestive and nervous systems. And so, let's meet these tiniest of wasps.

Chalcids often bear little resemblance to their larger relatives. They are all classified in a large, complex "superfamily" called Chalcidoidea, which includes 16 families. We're only going to meet a few; I will use the term "chalcid" here to refer to any of these wasps.

Chalcids attack a great variety of other insects, but like most other parasitoid wasps, they are often highly host-specific. They attack a great variety of

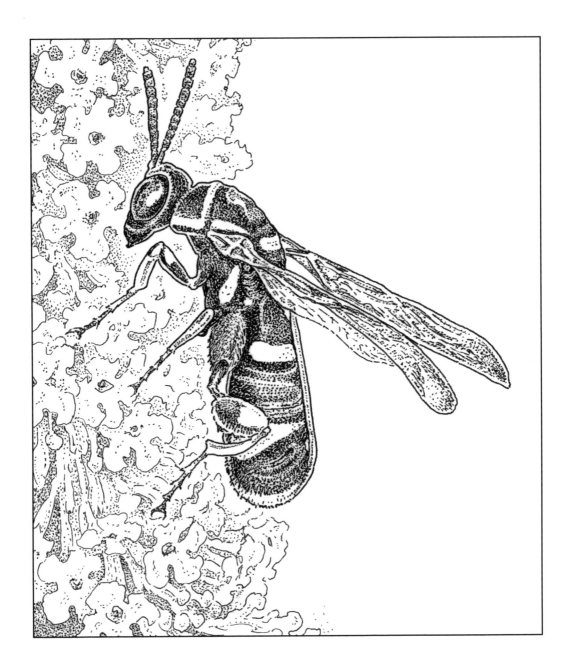

Leucospidae sp. on flower

other insects. When they target butterflies, moths, and other lepidoptera, they usually attack pupae. When they target flies, they are after the mature larvae.

Most adult chalcids seek nectar from flowers, but some feed on honeydew (the sugary secretion of aphids, scale insects, and related species). They are also attracted to nectar-producing glands elsewhere on some plants (called extra-floral nectaries, or EFNs), including on fruit trees such as peaches, cherries, and nectarines, and vegetable-garden stalwarts like beans, squash, and melons. They occasionally eat pollen. Because they are so small, they prefer the smallest flowers with easily available nectar. In addition to the ever-useful umbelliferous flowers, plants in the aster and mint families with small blooms also support them.

Stout, round-bottomed little wasps (family Leucospidae) make a brief appearance in my garden for a few weeks every summer, and are a welcome part of my yard's ecosystem. They resemble a small yellow jacket, until you look closer and notice they have something unique—a long ovipositor that curves upward over the abdomen, reaching to the thorax, giving this wee creature a unique and pleasing rounded look. These are somewhat rare, but you may spot them visiting flowers. Mainly parasitoids of solitary bees, they lay their eggs singly and externally. The first larva, once hatched, searches the host nest for any competitors, which are promptly dispatched. There is only one survivor, which develops externally (as an ectoparasitoid) on the host.

Among the smallest insects, so tiny you can barely see them, trichogramma-tids (family Trichogrammidae) are moved through the air by the slightest wisp of a breeze. Their miniscule wings are shaped like small paddles with a generous fringe. As you might expect, they are very weak flyers, but they make up for their small size in effectiveness. They are parasitoids of the eggs of many other insects, especially bugs, moths, butterflies, beetles, thrips, flies, and more. Many are used as biological control agents in agricultural and commercial settings.

Trichogramma sp. on egg

If you think the trichogrammatids are small, meet the enchanting fairy-flies (family Mymaridae), which are not flies at all but minute wasps. These insects are as small as the period at the end of this sentence. At $5/1000$ of an inch (0.15 mm), a mymarid takes the prize for smallest flying insect; they are among the most common chalcid wasps. Their adult lifespan, however, is very short, only a few days. During that time, they can attack and dispatch the eggs of some significant pests, including leafhoppers in rice crops and—of interest to us gardeners—eggs of true bugs, moths, butterflies, thrips, flies, and some other insects.

One of the largest and most versatile chalcid families, Pteromalidae, encompasses every form of parasitism we have discussed. Some are even predators. They attack a wide variety of hosts. Adults of some pteromalids puncture the host with their ovipositor and feed on their bodily fluids. Many are important biological control agents in agriculture, especially for some hard-to-control insects such as leaf miners.

APPEARANCE Adults mostly tiny, 0.1–0.75 in. (3–20 mm). All slender-bodied with narrow waists. Often metallic, black, dark blue, or green, but may be yellow and black. Chalcididae, 0.1–0.33 in. (3–8 mm), Leucospidae, 0.15–0.66 in. (4–17 mm), Trichogrammatidae, 0.01–0.66 in. (0.3–17 mm), Mymaridae, 0.01–0.04 in. (0.2–1 mm), Pteromalidae, 0.04–1.75+ in. (1–44 mm).

POPULAR NAMES Most have no common names; an exception is the fairyflies (Mymaridae).

SCIENTIFIC NAMES Superfamily Chalcidoidea, families Chalcididae, Leucospidae, Trichogrammatidae, Mymaridae, Pteromalidae.

COMMON TAXA With a few exceptions, most species of Chalcidoidea are unnoticed by gardeners.

DISTRIBUTION Chalcidoidea includes 2000 species in 18 families in America north of Mexico. The families discussed here include: 140 Chalcididae, 6 Leucospidae, 135 Trichogrammatidae, 202 Mymaridae, 600 Pteromalidae.

GARDEN ACTIVITIES Parasitoids of a broad range of insects and other arthropods. Some of the tiniest are the egg parasitoids. Fig wasp (family Agaonidae) larvae pollinate, then feed inside dioecious figs.

4

MEET THE
BEETLES

It may amaze you to learn that over one-fifth of all the living species on earth are beetles. But it probably doesn't come as a surprise that many can be found right in your garden. But first, let's clarify. What is a beetle, exactly?

Beetles (order Coleoptera) share two definitive features. They have hardened or leathery forewings (elytra), with membranous flight wings tucked under this protective cover. You may have observed these two wing types on lady beetles (the correct name for ladybugs) as they get ready to fly away. Many beetles can fly. Secondly, beetles have chewing mouthparts. Most eat plants, that is, foliage, fruit, seeds, and also wood. But quite a few eat other arthropods, including herbivorous insects. Some eat dung or fungus. Some are detritivores and even eat dead animals. Grubs (beetle larvae) may have different food preferences than adults.

Did you know some beetles act as pollinators? Among the earliest insects evolutionarily, beetles were here long before flowering plants and were among the first pollinators. They were first associated with cycads, then later with the early flowering plants such as magnolias and spicebush. Beetle pollination is known as "cantharophily," appropriately named after the flower-loving cantharid beetles, which are perhaps the ones most often noticed on blossoms. While the magnolia flower's sweet scent attracts everything from bees to flies, they are collecting nectar. Beetles, looking for pollen to eat, get the credit for pollination!

Providing suitable habitat for beetles is easy, but requires some thought beyond planting pretty flowers. Plants with clusters of small flowers attract pollen-feeding beetles. Many are particularly drawn to late-summer bloomers such as goldenrod and yarrow; they also like milkweeds and the umbellifers touted in earlier chapters. Early-blooming grasses are another important source of pollen for them. Beneficial beetles thrive in less-tidy gardens. Many live in leaf litter for parts of their life cycle, so leave them some unraked areas. Large stones, sections of logs set in borders, and even small piles of leaves or conifer needles also offer habitat. If you have a compost pile, chances are there are beetles already living and hunting in it.

Ever hear of beetle banks? These are intended to provide habitat for predaceous beetles and other wildlife. They're simply a wide berm of soil, raised 12 to 18 inches and planted with perennial grasses and non-woody (herbaceous) plants. They first appeared on farms in England, as a project of the Game Conservancy Trust. In a garden setting, it is sufficient to add "beetle bumps," generous mounded islands of perennial grasses mixed with flowering plants at least two or three feet in diameter and, like the banks, 12 to 18 inches high. Such a mound keeps ground-dwellers and overwintering beetles dry in winter, while the plant cover moderates temperature swings.

There are so many beetles, I will barely be able to scratch the surface of the great numbers of welcome garden species, but will survey the most common taxa in the next few essays. Keep your eyes open for all of these and more when you are out in your garden.

The Voracious Lady Beetles

Lady beetles' bright coloration warns predators to leave them alone. And, just in case an unwary bird decides to have a taste, they excrete a noxious-tasting fluid from their leg joints.

There are those who don't really like insects, but it's hard to find anyone who doesn't have a soft spot for lady beetles (family Coccinellidae). Lady-bugs, as they are more familiarly called, appear in legends, songs, and children's stories, and are a popular decorative motif. Beloved natural-history author Sue Hubbell whimsically dubbed them "the panda of the insect world." It is easy to believe that the phrase "cute as a bug" is a reference to the cheery polka-dotted beetle. But while even small children can recognize a typical lady beetle, very few of us really know much about this gardener's good friend.

Lady beetles vary in size, ranging from the unexpectedly small to the bigger species we are used to seeing. They even come in different colors. In fact, I have not yet identified all the species I have found in my own garden! We all recognize the round, dome-shaped, and spotted "classic ladybug." You might also see the oval and flatter-bodied convergent lady beetle. The parenthesis beetle is named for the distinctive marks on its forewings (elytra). The aptly named seven-spotted lady beetle is a European species, first introduced to North American for biological control in the 1950s. My favorite, not often seen, is the dramatically named "twice-stabbed," an elegantly shaped dome of shiny black lacquer, splashed with matching red spots on its forewings. Most lady beetle species are about an eighth of an inch across. However, once

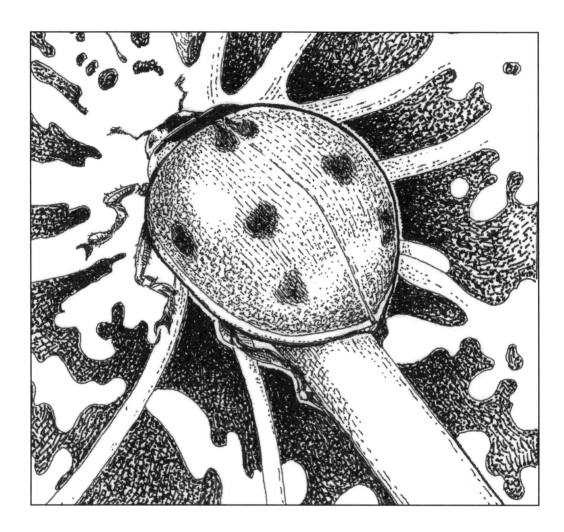

Seven-spotted lady beetle (*Coccinella septempunctata*)

emerged from the pupa, they stop growing (as do all adult insects that undergo complete metamorphosis). Therefore the diminutive pale orange ladybugs you may come across are not juveniles, but adults of smaller species. Typically, lady beetles' life cycle is four to seven weeks; some species, however, may live up to three years.

The appetite of lady beetles is nothing short of prodigious. While some eat a more general diet, snacking on a variety of soft-bodied insects, many are specialist predators, eating only certain other species of insects. The good news for us gardeners (and farmers) is that as both larvae and adults, they eat many common pests. Depending on the species, a single larva may consume 350 aphids and an adult can hoover up to 5000 aphids in a lifetime. It is fascinating to watch them voraciously cleaning up a colony of aphids. Scale-eating species (scale is another tiny, soft-bodied, plant-eating insect pest) can be equally effective, but since in that case the lady beetle larvae feed underneath the scales, we don't see them at work. Other lady beetles specialize on mealybugs, whitefly, mites, and insect eggs.

The modern science of "classical" biological control, that of using a non-native predator to control a non-native pest from the same region, was pioneered by Charles Valentine Riley, an entomologist with the USDA and the first curator of insects at the Smithsonian Institute. In 1888, he had the game-changing idea to search for natural enemies of an exotic pest, the cottony cushion scale, on its home ground in Australia. He successfully introduced vedalia beetles (a species of lady beetle), along with a parasitic fly, both from Australia, to California citrus orchards. He was thus credited with saving the local citrus industry in less than one year! Other lady beetles, such as the "mealybug destroyer," have also proven useful in commercial pest management.

And in gardens? If we have an aphid or mealybug problem, can we deploy lady beetles? You may have seen mesh bags of lady beetles for sale (usually *Hippodamia convergens*, a West coast native migratory beetle) for release into

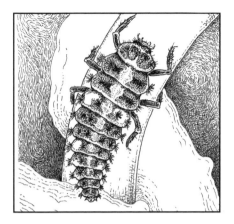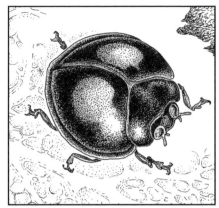

Left: Larva of convergent lady beetle (*Hippodamia convergens*)

Right: Twice-stabbed lady beetle (*Chilocorus kuwanae*)

the home landscape. This tack is often unreliable and always ecologically unwise. Beetles collected from overwintering aggregations disperse quickly when released in a garden, rarely stopping to feed. Beetles collected later, in late spring, may not disperse as quickly but they also will not eat much prey, as they have high fat reserves. Besides, that black "confetti" you see underneath the rack of mesh bags full of ladybugs for sale? Ladybug legs. In other words, sad to say, though alive, the helpers are hobbled.

How, then, can we encourage this useful predator to arrive and stay in our gardens? Here are a couple of useful tips. Provide habitat. Not all species of lady beetles migrate, and even migratory species may form year-round populations in mild-winter areas. The shelter of leaf mulch, perennial bunch grasses, and groundcovers will give them safe spots to overwinter. There they may form clusters (aggregations) of anywhere from hundreds to thousands of individuals.

Here is a good example of why to take a food-web approach to your garden: by including plants like sunflowers, milkweed, and German chamomile,

Convergent lady beetle (*Hippodamia convergens*)

which aphids and other soft-bodied insects favor, we provide an alternative food source when there are none of these pests on our rosebuds, broccoli, and other valued plants.

Remember, they are omnivorous, so offer lady beetles season-long nectar and pollen sources, which they require during parts of their life cycle. In early spring, I always have thriving populations of lady beetles on the common reseeding annual German chamomile—it functions as a virtual lady beetle nursery! Other plants that will sustain them include cosmos, coreopsis, alyssum, and umbelliferous herbs such as parsley, dill, cilantro, and fennel.

It is not hard to attract and support many different species of lady beetles. Do not buy them! Luring them with an attractive and varied garden is more effective and cheaper than fighting bothersome insects with pesticides. Who doesn't prefer spending money on beautiful plants instead?

APPEARANCE Oval yellow eggs, sometimes in clusters, often near or in aphid colonies. Larvae resemble miniscule snub-nosed alligators. Odd-looking rounded pupae are often attached to plants. Adults, 0.03–0.4 in. (1–10 mm), oval or round, dome-shaped; elytra various shades of orange or bright red, but may be pale yellow, or black. May have spots or other markings.

POPULAR NAMES Ladybugs, lady beetles, lady bird beetles.

SCIENTIFIC NAMES Order Coleoptera, family Coccinellidae.

COMMON TAXA Convergent, parentheses, two-spotted, and twice-stabbed lady beetles. Tiny spider mite destroyers are rarely noticed, the ashy grey lady beetle is arboreal, and the introduced Asian multicolored lady beetle (a pest in some regions, and at some times of year) exhibits a variety of spotted patterns.

DISTRIBUTION 480 species in America north of Mexico; includes species introduced for biological control.

GARDEN ACTIVITIES They eat a variety of small soft-bodied garden insect pests; specialists feed on only one or several related species of prey. (There are a few herbivorous lady beetles; some are pests of cucurbits and beans.)

Soldier Beetles
on the March

No, they aren't blister beetles! Several insect species, including soldier beetles, closely mimic the appearance of the related blister beetles, taking advantage of the blister beetles' chemical defenses as protection from predators.

In his detailed and fascinating 1964 book *Beneficial Insects*, Lester Swan comments on several beneficial species of soldier beetles (family Cantharidae) and their associated prey, then notes that "unfortunately, they have not been studied extensively." Not much has changed in the last few decades. "Oh, those! I have them in my garden, but I didn't know they were beneficial," is the familiar refrain when I point them out to garden visitors. Yet soldier beetles surely warrant the same recognition given to lady beetles and lacewings. They are valuable and reliable allies, and find amenable living conditions in gardens more easily than in agricultural fields. Their lack of potential for commercial use explains why soldier beetles have been so little studied.

Soldier beetles have a voracious appetite for aphids, caterpillars, grasshopper eggs, mites, and other small bugs. They are even reputed to eat cucumber beetle eggs, reason enough for any vegetable gardener to agree that they deserve further study!

Let me describe soldier beetles—you might recognize them. The adult is a narrow, parallel-sided beetle, with long threadlike antennae. Most species have dark heads, a red or orange segment (the pronotum), and dusty-looking gray, brown, or sometimes bluish wing covers (elytra). The wing covers sometimes

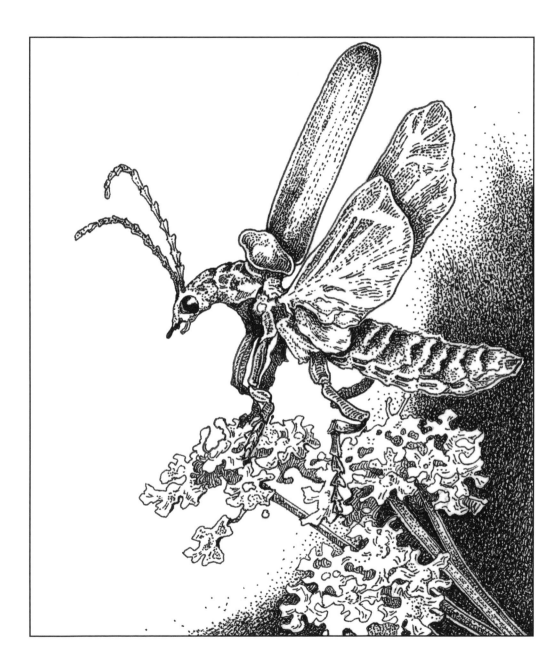

Cantharis sp. taking flight

have markings and, unlike as in most beetles, are relatively soft, resulting in a second common moniker, leatherwings.

These beetles may be mistaken for other, similar-looking related beetles, so check your identification. Unwelcome in gardens, blister beetles can raise blisters on any unfortunate gardener who touches them. Click beetles are well-known to children, who like to place them on their backs and hear the satisfying "click" as they flip over. Adult male glowworms look similar but have long, feathery antennae.

A female soldier beetle sometimes attracts hordes of males with the pheromones she emits, but generally only one male is successful. Most beetles do not engage in elaborate courtship behaviors, but some soldier beetle males may "nibble" females. When these beetles appear in my garden, there seems to be a lot of nibbling going on. I have to wonder what attracted the female to the lucky winner!

Each female has a huge supply of eggs, which hatch in spring. The larvae are nocturnal and can be quite long-lived, staying in this stage for up to three years in some species. You might come upon them in damp areas beneath rocks, or nestling in leaf litter and under bark, where they hunt for insects and other small organisms. A few species are minor pests as larvae, when they feed on roots. Once they pupate and emerge as adults, they live less than a year.

Many are primarily carnivorous in both the larval and adult stage. (An exception: the genus *Chauliognathus* are pollen feeders as adults.) Because they are generalist predators, they sometimes eat other desirable insects such as lacewing larvae as well as aphids that have been parasitized by wasps. Larvae consume eggs and larvae of other beetles, moths, grasshoppers, and more. Adults are frequently found on a variety of flowers, where they feed on pollen and nectar in addition to prey such as aphids and mealybugs.

A frequently seen soldier beetle in the West, the adult brown leatherwing beetle (*Pacificanthia consors*), helpfully dines on citrus mealybugs. The

Pacificanthia consors

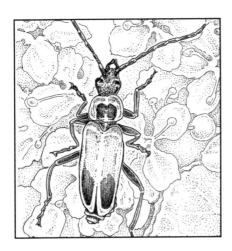

Chauliognathus pensylvanicus

Pennsylvania leatherwing or goldenrod soldier beetle (*Chauliognathus pensylvanicus*) is common in the Midwest and East. Its larvae begin their lives in soil, where they prey on small soft-bodied insects, grasshopper eggs, and small caterpillars. The adults are often spotted in great numbers in late summer on goldenrod and Queen Anne's lace, dining on insects, nectar, and pollen and looking for mates.

Encouraging a large garden population of soldier beetles is simple. It is important to their life cycle (as well as that of other beneficial organisms) that they have undisturbed, moist mulched soil in which to pupate, so leave unraked areas. Add organic matter to the surface of your perennial beds as needed to keep the soil in good fertility. Remember, in addition to soldier beetles, a fragile and important community thrives at the interface between organic matter and the soil below. Each spring, I look forward to the day when I spot the first soldier beetles, knowing that soon there will be an abundance of these helpful allies patrolling my garden.

APPEARANCE Usually 0.5 in. (13 mm) or smaller, narrow, parallel-sided. Often with dark head, and reddish pronotum (plate covering thorax) and velvety soft, bluish, brown, or gray elytra.

POPULAR NAMES Soldier beetle, leatherwings.

SCIENTIFIC NAMES Order Coleoptera, superfamily Elateroidea, family Cantharidae.

COMMON TAXA Species of the following genera are commonly seen in gardens, depending on region: *Cantharis*, *Podabrus*, *Chauliognathus*, *Rhagonycha*, and *Pacificanthia*.

DISTRIBUTION In America north of Mexico, 470 species. Many species of soldier beetles are widely distributed.

GARDEN ACTIVITIES Larvae are mostly carnivorous, feeding on small soil organisms; a few feed on roots of grasses, potatoes, and celery. Adults of many species are important predators of aphids. Adults also eat nectar and pollen, and often other insects such as mealybugs, small caterpillars, and other soft-bodied insects.

Predaceous Ground Beetles on the Prowl

Unlike the majority of its grounded kin, the fiery searcher beetle (also known as the caterpillar hunter) sometimes climbs trees and shrubs in search of its favorite prey.

While abundant in most gardens, we rarely notice predaceous ground beetles (family Carabidae) thanks to their primarily nocturnal habits, dark colors, and quick movements. The carabids are one of the largest families of beetles and hunt caterpillars, beetle grubs, grasshoppers, and other small invertebrates such as snails and slugs. Very few are plant-eaters and they are only incidentally spotted on flowers. They have prodigious appetites and can eat their weight in prey daily! It is suspected that these carnivores may provide significant biological control in agriculture, but the science is sparse.

The carabids have streamlined, somewhat flattened bodies and long legs that help them move quickly and distinguish them from sometimes similar-appearing but slow-moving common darkling beetles (family Tenebrionidae). Also, the herbivorous darkling beetles are rounder and have antennae that look like a string of beads, or are enlarged at the ends, while ground beetles have threadlike antennae. Some species have fused forewings (elytra) and are flightless. Perhaps the easiest way to identify ground beetles is to watch them. As a rule, any hunting beetles are faster and more nimble than herbivorous species.

Both carabid grubs and adults primarily hunt and feed in soil and leaf litter. They often shelter at the base of perennial bunch grasses and low-growing groundcovers. They are one of the most frequently found beetles in beetle

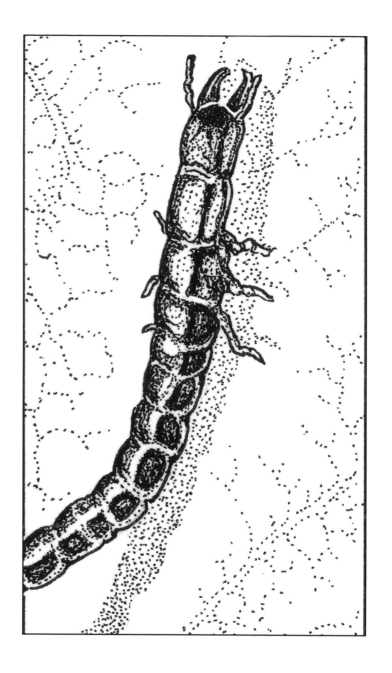

Ground beetle larva (*Pterostichus* sp.)

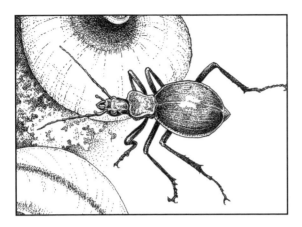

Left: Snail-killer carabid (*Scaphinotus* sp.)

banks and bumps (as previously described). Most emerge from their hiding places at dusk to swiftly run down their prey.

The most common ground beetles seen in our yards are the easily recognized *Calosoma* species. Grubs (beetle larvae) may take a year to develop; they have dark segmented bodies that superficially resemble elongated roly-polys. Adults, which can live up to three years, are dark-colored (sometimes metallic) and usually no more ¾ of an inch long. *Pterostichus* is a very large genus of ground beetles that includes many garden species. They are generally predators as larvae and adults but, like us, eat the occasional strawberry. And if you host *Scaphinothus*, you are a lucky gardener—they prey on snails.

Other interesting species of ground beetles are more often encountered in the wild than in gardens. You may have heard of bombardier beetles, or even watched entertaining videos of their behavior on YouTube. When threatened, they aim and forcefully eject a stream of boiling, noxious liquid that vaporizes with an audible pop as it contacts air. Related species share the ability to repel enemies by releasing noxious fluids, but none with the spectacular show of the

Ground beetle (Pterostichus sp.)

bombardier beetle. Beautifully marked tiger beetles, which often hunt during the day, are found along sandy riverbanks and other sandy areas.

While carabids in general are common beetles, some are highly restricted in distribution. For instance, viewing the critically endangered beautiful Delta green ground beetle (*Elaphrus viridis*) requires a journey to Solano County, California, where it is found only at the margins of vernal pools on the Jepson Prairie Preserve.

Although ground beetles are impressive carnivores, there are, as always, some exceptions. A few eat decomposing organic materials and, as grubs, eat roots. Some ground beetles consume seeds (which is considered a form of predation because ingesting a seed kills an organism; it will never sprout). In my own garden, I have turned over pots and rocks and startled small copper-colored scurrying ovals (sometimes black or brown), seed-eaters in the genus *Amara*. They quickly dart about when a flash of light illuminates them, as does another taxon of seed-feeders, the *Harpalini*. I have not found any of these seed-eating ground beetles to cause any garden problems.

It is easy to encourage the presence of ground beetles in your garden. As ever, remember to include a diverse array of plant species, which will attract beetle prey. In addition to perennial grasses, provide abundant dark and damp hiding places such as rocks, stepping stones, potted plants, and a few logs and branches, so they can be safe from their own predators (which include toads, ground-feeding birds, moles, and more). As dusk falls, I am delighted to know that the vast family of carabids is emerging for the nightly hunt.

APPEARANCE Size ranges greatly; most 0.1–1.2 in. (3–30 mm). Often shiny; they are usually dark colored, often black, but may be brown or reddish. Some are iridescent and brightly colored, especially green or purple shades. Larvae have large heads with visible mandibles.

POPULAR NAMES Ground beetles, predaceous ground beetles.

SCIENTIFIC NAMES Order Coleoptera, family Carabidae, which includes subfamily Cicindelinae, formerly its own family.

COMMON TAXA Genera such as *Calosoma*, *Pterostichus*, *Agonum*, *Harpalus*, *Lebia*, and *Stenolophus* include common species of ground beetles. Seed-eating ground beetles (*Amara* and *Harpalini* spp.), rufous carabid (*Calathus ruficollis*). Snail eater (*Scaphinotus* spp.) may be spotted by lucky gardeners.

DISTRIBUTION Over 40,000 species worldwide; over 2000 species in the United States.

GARDEN ACTIVITIES Most are generalist predators in both grub and adult stage, feeding on insects and other invertebrates in soil and leaf litter. Some feed on snails and slugs. A few eat seeds and decomposing organic materials.

So Many Rove Beetles!

A great number of species in the enormous family of rove beetles is present in the average garden, but often hidden from sight. Both the larvae and the adults of most species prey on all kinds of insects, mainly in soil and in leaf litter.

In this chapter's introduction, I related that over one-fifth of all the living species on earth are beetles. Until recently, entomologists had determined that there were more weevils (family Curculionidae) than any other beetle. But as exploring and classifying continues, we find that rove beetles (family Staphylinidae)—over 63,000 described species and counting—are now at the head of the pack. Even so, few have been studied. Once again, I suspect the lack of research may have to do with a perception that rove beetles are not useful in agricultural systems (though some are now being studied in Canada and Europe). Let's get to know them.

Rove beetles are usually easily recognized by their short forewings (elytra), which leave over half their abdomen exposed. A few other beetles also have truncated forewings, but these tend to have hind wings peeking out untidily, while those of the rove beetles are neatly folded underneath. These wings can take them to the air, though their usual habit is to run around on the ground. To the untutored eye, rove beetles resemble earwigs, but note the absence of pincers. Adult rove beetles have large mandibles that they use to grasp prey. In this diverse family, there is a lot of variation, from minute forms to species as big as an inch and a half, but most are in the range of a mere third of an inch. They may be brown, black, or red, or combinations of these colors.

Devil's coach horse (*Ocypus olens*)

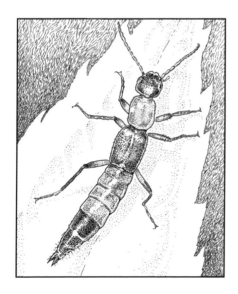

Paederus sp.

Rove beetles eat just about anything, except (with very few exceptions) living plants. The majority are predatory. Many have prey preferences, but are also opportunistic (the term for this behavior is "facultative predation"). Their diet consists mainly of insect larvae and pupae and other small invertebrates. Some are scavengers rather than hunters. A few species are parasitoids, an uncommon life strategy among the beetles. Others have a mutualistic relationship with mammals, eating parasites such as fleas, undoubtedly to the great relief of the mammal hosting the rove beetles. A few nibble on roots or on rotting vegetation. Adults of some species are even pollinators. In the mountains of the Pacific Coast, *Pelecomalium testaceum* is a pollinator of yellow skunk cabbage.

The hairy rove beetle (*Creophilus maxillosus*), in both its larval and adult stages, feeds on fly maggots and may be helpful in keeping down populations of pesky flies. It can be found on road kill and other animal carcasses, or in association with dung, all great places to find fly larvae, of course.

The devil's coach horse (*Ocypus olens*), an introduced species, can exceed an inch in size. It is eyebrow-raising when first spotted, as it elevates its long abdomen and opens its jaws to mimic a scorpion when threatened. It not only has a painful bite, but also a defensive secretion to repel predators. In both its larval and adult stages, it consumes many insects as well as earthworms.

Many other rove beetle species curl their abdomen over their thorax as a threat, and defensive secretions are common. Some members of the genus *Paederus* produce a highly toxic skin irritant, pederin, that can result in a painful blister. Having once caught one of these beetles in a slip of folded paper that I slid into my pocket, I was lucky never to learn the location of the escapee! Pederin is a contender for the most powerful animal toxin. Reputedly more powerful than cobra venom, it can do serious damage to human skin and eyes. Interestingly, it is being investigated for its medicinal properties and shows promise in eradicating some cancerous growths.

Rove beetles can be found in just about any habitat, but always where there is moisture. Look for them on decomposing fruit and under the bark of decayed trees. Many are specialized. Hundreds of species live only on the seashore, and some soil-inhabiting species no longer have the eyes they don't need. Other species live in the nests of social insects, doing no harm but providing no benefit ("inquilines"). Others climb on plants to hunt, especially at night.

Rove beetles of all sizes are easily accommodated in a garden. They benefit from beetle banks, and are yet another insect that thrives when soil and leaf litter is left undisturbed. In addition to leaf litter, they overwinter in compost and rock piles, or anywhere there is a damp and safe hiding place. Provide suitable habitat and rove beetles will move into your garden.

APPEARANCE Size varies, but most are under an inch (8 mm). Generally elongated, with truncated elytra. Colors range from yellow and red to brown and black. A few are iridescent blue and green. Some run with their abdomen curled over their thorax.

POPULAR NAMES Rove beetles, devil's coach horse.

SCIENTIFIC NAMES Order Coleoptera, family Staphylinidae.

COMMON TAXA Devil's coach horse (*Ocypus olens*) is introduced. *Paederus* species.

DISTRIBUTION 4400 species (and many more undescribed—it's estimated that there are over 5000 in America north of Mexico). About 63,000 species worldwide; probably over 70 percent of tropical species remain undescribed.

GARDEN ACTIVITIES These beetles will eat just about anything!

Flower, Dung & Other Beneficial Beetles

You can easily recognize tumbling flower beetles. Just brush against some garden blooms, and you will immediately see how these little acrobats acquired their name. Many are decomposers, but a few prey on insects.

Dung beetles, a type of scarab (family Scarabaeidae), were regarded as sacred by the ancient Egyptians, who perceived in the insects' efforts a symbol of the sun crossing the sky. Our contemporary appreciation of the dung beetle is more prosaic; because it efficiently rolls up and buries balls of manure (on which it lays its eggs), some species have been imported to control flies in rangelands and pastures in some of the western and southern states. Most of our native North American dung beetles are "dwellers," with larvae living in the animal pat, and "tunnelers," pulling manure into the soil directly under the animal pat. Only a few of the native North American species are the entertaining rollers, gathering manure and transporting it to their nest. They all have important roles to play, as without the scarabs dealing with the piles of manure, we would be deep in it!

Scarabs include many pest species, including the common and pesky June bugs (*Phyllophaga* spp.) and the annoying Japanese beetles (*Popillia japonica*). Their grubs feed on roots, while the adults feed on fruit. They range greatly in size, but can be recognized by their distinctive platelike antennae. They also tend to be stout, with fuzzy abdomens; the front legs of many species are adapted for digging. The grubs have a characteristic "C" shape. Some species are metallic, resembling molten gold, silver, or bronze, and are prized by

Dung beetle (Scarabaeidae)

collectors for their great beauty. Most adults are nocturnal, and they are often attracted to lights. Individuals may live for up to four years.

The checkered beetles (family Cleridae), however, are welcome garden residents. They are long and lean, and clothed in short hairs. They can attain an inch in length, but are usually much smaller. Nonetheless checkered beetles such as *Tanaops longiceps* are conspicuous because of their brightly checkered patterns of red, orange, yellow, and sometimes blue coloration. Predaceous in both the larval and adult stages, they are notable for their voracious appetite and particularly for eating other beetles, in addition to other small insects. Some species are scavengers, found on carrion. Many species visit flowers, where the adults feed on insects and pollen. Some clerids are even found up in trees, where they burrow into the bark, feasting on insects that dwell there; these may yet prove useful in our efforts to control bark beetles.

A few species of checkered beetles are "nest robbers," feeding on larval bees, wasps, and termites. Their larvae most often eat eggs and larvae, while adults prey on adults. The ornate checkered beetle (*Trichodes ornatus*) begins life as a kleptoparasite in the nests of ground-nesting leafcutting and mason bees, where it feeds on both pollen and bee larvae. As an adult, it eats pollen, but is not averse to catching a passing bee. Look for adults on yellow flowers, as well as flowers of yarrow, buckwheat, and ceanothus.

Closely related to the clerids, soft-winged flower beetles (family Melyridae) are often found in our yards; they sport similar form and coloring. However, they are generally smaller; some species can be about a half-inch, but more usually, they are less than half that. Their bodies have a soft, plump appearance. Most adults feed on flower visitors and pollen. In North America, *Collops* species are the most common garden visitors; they are black with orange markings, sometimes appearing as an orange cross on their back. The predaceous larvae are usually found in soil or leaf litter, and sometimes under bark.

Tanaops longiceps (Melyridae)

Eustenopus villosus (Curculionidae)

Peer into flowers frequently enough (I hope you do!) and you are bound to delight in the tumbling flower beetles (family Mordellidae). Only a few prey on other insects, but they serve as decomposers. They are, as their name suggests, usually spotted in flowers.

Most gardeners do not consider weevils (family Curculionidae) to be beneficial and indeed many are quite destructive. But some of them are highly host-specific and they can be effective at controlling certain invasive weeds, attracting research interest. A challenge in these situations is finding out if they will stick to feeding on the intended target and not move on to other hosts (a phenomenon known as "host switching"). In the eastern United States, an introduced weevil exclusively attacks the rampant mile-a-minute vine. In Minnesota, both seed head and root weevils are cost-effective controls for spotted knapweed. The USDA recently approved the use of the weevil *Eustenopus villosus* to control invasive yellow star thistle. Some species of leaf beetles (family Chrysomelidae) and the related tiny seed beetles (family Chrysomelidae, subfamily Bruchinae) are also used to control noxious weeds. Suffice it to say, there is potential here for effective weed control.

Did you know that fireflies or lightning bugs (family Lampyridae) are beetles? And that we should welcome them to our gardens for more reasons than the magical pleasure of viewing their winking light shows on summer evenings? A common sight east of the Rockies, they are increasingly threatened by habitat loss, pesticide use, and possibly light pollution. When I was a child and my father was away conducting research in the jungles of Malaysia, he sent home a holiday letter with a memorable story that featured the synchronous flashing of the fireflies he saw there decorating a "Christmas tree." Synchronously flashing fireflies are also found in the Great Smoky Mountains, but most species have a unique pattern of flashes to attract mates, a trait that is occasionally taken advantage of by mimics.

Fireflies are garden-friendly beetles. But adults are not the only ones that glow. In some firefly species, both the eggs and the larvae emit a soft glow, though because they are soil-dwellers we may not see this. The larvae, dubbed "glowworms," do look like small segmented worms, albeit with flattish overlapping scales, no more than an inch long. Specialized predators of soft-bodied garden pests, they consume mites and mosquito larvae as well as snails and slugs. The short-lived (a couple of weeks is a typical span) adults may feed on pollen and nectar. How can you encourage them? Do not use pesticides. Turn off outside lights at night. Mow your lawn less often, especially near trees. The females will be watching from the branches when the males start winking in the taller grass below.

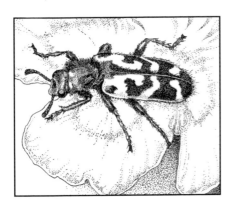

Ornate checkered beetle (Cleridae)

APPEARANCE Color varies greatly. Size varies greatly, too, from 0.03 in. (1 mm; Chrysomelidae) to 2.5 in. (63 mm; Scarabaeidae).

POPULAR NAMES Scarabs, checkered beetles, soft-winged flower beetles, tumbling flower beetles, weevils, leaf beetles, seed beetles, fireflies.

SCIENTIFIC NAMES Families Scarabaeidae, Cleridae, Melyridae, Curculionidae, Chrysomelidae, subfamily Bruchinae, Lampyridae.

COMMON TAXA Scarabs are fairly common in gardens; fig beetles, June beetles, rose chafers, Japanese beetles, and Hercules beetles are all scarabs.

DISTRIBUTION Species in America north of Mexico discussed here: 1400 Scarabaeidae, 300 Cleridae, 520 Melyridae, 2500 Curculionidae, 1900 Chrysomelidae, 150 Bruchinae, 170 Lampyridae.

GARDEN ACTIVITIES Many beetle families include both predaceous species and scavengers. Some are plant feeders, welcome if they feed on unwanted weeds.

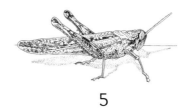

5

THE GARDEN COMMONS

True Bugs, Lacewings & Other Familiar Garden Insects

This chapter represents a broad range of the biodiversity of the insects, including some you probably have heard of or recognize. But what exactly is biodiversity? For a gardener's purposes—no matter how big or small your patch of land is—it is important to have a broad range of creatures (species richness). But it is also important to have not only a lot of species, but also enough individuals of each kind (abundance). Furthermore, you want different species exploiting the same resources (in the same biological guild). For example, it's good to have lady beetles, parasitoid wasps, *and* lacewings in residence to regulate the aphid population. All of these things together—the number of species, the number of individuals of each species, which species they are, and what they are doing—all add up to something we call "functional biodiversity." This leads to a healthy garden ecosystem, boosting its resilience (its ability to bounce back from disturbances). A garden with resiliency can surf the waves of disturbance, such as planting new beds or pruning, or an invasion of aphids, and even the natural disturbances of seasonal changes, as they occur.

As a rule, the higher the number of plant species in your garden, the higher the number of arthropod species it will support. A biodiverse landscape full of insects fosters "insurance species" of insects, those in the same guild (for example, the lady beetles, parasitoid wasps, and lacewings that all prey on the same food). These are "understudies" that are ready to step into a role when another species, for whatever reason, becomes unavailable. Yes, some of the following creatures can be perceived as pests, but they are an essential part of a garden ecosystem, as you will discover!

Many "true bugs" are prominently featured in this chapter. As I commented back in this book's introduction, when it comes to the world of small creeping, flying animals, we often use "bug" as a catch-all. But entomologically "bug" has a specific meaning—it refers to just one order of insects, the Hemiptera, and more specifically to the suborder Heteroptera, distinguished from all other insects by the common name "true bug." Hemiptera are characterized by their unique piercing, sucking mouthparts. In some cases, those mouthparts are used for piercing plants, in some cases other animals (prey), and sometimes also for defense.

Heteroptera means "different-winged," and in this taxon, only the tips of the first pair of wings are membranous, while the second pair is fully membranous. Also, fascinating to behold, these insects are often masters at the art of hiding by camouflage, blending into background scenery. Some are able to masquerade as an inanimate object of no interest to predators or prey, such as a leaf. They also commonly engage in mimicry, looking like other insects such as bees, wasps, or ants. Some are adorned with spines or other forms of physical defense; many also have chemical defenses. Our gardens wouldn't be the same without these intriguing insects.

The first two essays will introduce a variety of common heteropterans, including some valuable predators. After that, we'll explore the true bugs that were once in a suborder called Homoptera, "same-winged," but are now separated into two unrelated suborders. You'll recognize many of these.

I have included several other taxa in this chapter: grasshoppers, katydids, and crickets (Orthoptera), and the mantises (Mantodea), lacewings (Neuroptera) and ants (Hymenoptera). True, some of these can be injurious to plants, but others are useful predators. While it is not always possible to point to a specific or obvious benefit—from a human's perspective—remember that herbivorous insects that may damage our plants are food for something else. They are beneficial because they are part of the larger garden ecosystem. Ultimately, they do benefit us with the reward of a healthy garden, in which we can sit back and let the predators and parasitoids do most of the work of pest management. The key is to know when you need to intervene, and when you can step back and let nature take care of the problem.

In the long run, developing "functional biodiversity" in your garden decreases the likelihood of pest outbreaks that require attention, and allows you to focus on more rewarding aspects of gardening. As a bonus, you'll get to observe many interesting insects.

TRUE BUGS

Friends or Foes?

True bugs have piercing, sucking mouthparts, but they can be hard to see. They usually fold underneath the bug into a groove, much as a penknife when closed. The bugs extend their mouthparts to feed.

Some families of the "true bugs" are exclusively predaceous; we will meet them in the next essay. Here are some examples of heteropteran families often present in gardens that are herbivores, omnivores, and/or predators, depending on species. Some are food for other garden residents. Before jumping to any conclusions, carefully observe what they are doing. Many of the plant-feeding species are only present in small, innocuous numbers. In larger numbers, such as colonies of aphids, they may be hosting desirable predators.

First, you can confirm they are heteropterans if only the tips of their first pair of wings are membranous. Many have a triangle on their back (the "scutellum"). Next, notice their piercing, sucking mouthparts. When they damage plants, you'll find them difficult to control with a spray because those mouthparts bypass plant surfaces. However, you can usually exclude them from garden crops with row covers. Heteropterans are also distinguished by having scent glands. These are not always easily detected, although they are immediately evident to anyone who has bitten into a raspberry that was visited by the common stink bug!

Stink bugs (family Pentatomidae) can be tricky to categorize because some are herbivores and some are predators, and they can look very similar. The

Two-spotted stink bug (*Perillus bioculatus*)

common spined soldier bug (*Podisus*) preys on gypsy moth caterpillars, Colorado potato beetles, and Mexican bean beetle larvae—without a doubt, valuable services! Yet many of these have been summarily squished, a case of mistaken identity. Meanwhile, the non-native Asian brown marmorated stink bug (*Halyomorpha halys*), which lacks predators and parasitoids in this country, has become a pest of farms, gardens, and households. An easy way to tell them apart: the predatory spined solider bug has sharp spines on its back; the pests have blunt spines. Keep an eye out as well for the two-spotted stink bug (*Perillus*), which also consumes Colorado potato beetles.

The graceful and rarely noticed stilt bugs (family Berytidae) are mainly plant-feeders. Slender and long-legged, with swollen-looking joints, like most herbivorous insects, they tend to move slowly. Some gardeners and farmers consider them pests, as they feed on everything from geraniums to corn, tomatoes, okra, and peaches. But they are generally present in small numbers, and also dine on other slow-moving and immature insects, including aphids, caterpillars, and various insect eggs.

The diminutive plant bugs (family Miridae), with their long thin antennae and slender legs, are also primarily plant-feeders and common in most gardens. But this large family also includes many important predatory species known to eat aphids and mites.

We can simply regard true bugs such as these as part of the interdependent garden food web. They provide a good food source for other insects, birds, lizards, spiders, and other animals, even though some may do occasional damage to our plants. While there may sometimes be good reason for intervention, the best and simplest pest management strategy is a well-planned habitat garden, where all may live in relative harmony.

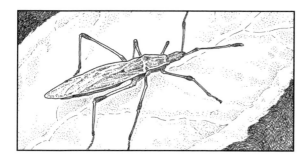

Stilt bug (Berytidae)

APPEARANCE Piercing, sucking mouthparts. Many heteropterans have a prominent triangular "scutellum" on the dorsal side. Size varies. Berytidae, 0.2–0.35 in. (5–9 mm); Miridae, 0.05–0.6 in. (2–15 mm); Pentatomidae, 0.2–0.7 in. (5–18 mm).

POPULAR NAMES Stink bugs, stilt bugs, plant bugs.

SCIENTIFIC NAMES Order Hemiptera, suborder Heteroptera, families Berytidae, Pentatomidae, Miridae.

COMMON TAXA Heteropteran families that are principally herbivorous, but include some predators: one stink bug family (Pentatomidae, subfamily Asopinae), includes spined soldier bugs. Knob-kneed stilt bugs (Berytidae). Plant bugs (Miridae), the largest heteropteran family.

DISTRIBUTION Most families include common, widely distributed garden species. In America north of Mexico, 220 Pentatomidae (of which 35 are subfamily Asopinae), 12 Berytidae, 3850 Miridae.

GARDEN ACTIVITIES Feed on a wide variety of plants, animals, and other insects—and serve as food for others.

Fierce Predators

Ambush bugs are such powerful predators that they are able to capture prey up to ten times—or even more—their own size, including the stinging hymenopterans.

The predatory true bugs are good friends to gardeners—and besides, with such wonderful common names, who wouldn't wish to invite them in? Families found in healthy garden ecosystems (all in the suborder Heteroptera, which we began exploring in the previous essay) include damsel bugs and assassin and ambush bugs. The big-eyed bugs and the insidious flower and minute pirate bugs, with few exceptions, are also heteropteran predators. Strong mouthparts and powerful raptorial front legs for clutching prey, such as the mantises (order Mantodea) have, are also common characteristics.

Ferocious and voracious, these bugs are great hunters. Among their many prey are arthropod eggs, aphids, mealybugs, scale insects, whiteflies, leaf hoppers, caterpillars, thrips, beetles, and more. They may also feed on other predators (this is called "intraguild predation"). Some are generalist predators, but many are specialists and have specific favored foods. While some are active hunters, many, such as ambush and assassin bugs, are sit-and-wait predators, pouncing on inattentive prey that wanders by.

Predatory bugs, like other predatory insects, are especially in need of our help. Because they are at the top of the insect food chain, they are at risk. If the prey they consume has been exposed to pesticides, the chemicals concentrate

Jagged ambush bug (*Phymata* sp.)

in their bodies. Birds and other larger predators of these insects may also be poisoned as a result. And when predators die, it unfortunately falls to us to manage their unchecked herbivorous insect prey. Don't use pesticides, please! As a bonus, you get to watch entomological theater unfold.

"Watch them, but don't pick them up" is a good general rule for any unidentified insect. I will never forget an incident from when I was in college. "Quick, grab it!" my roommate, enrolled in entomology, exclaimed when we spotted a very large bug, "It won't hurt you." Well, you probably know the end of that story—I learned the hard way to always identify insects (in this case, an assassin bug) before I pick them up! With their beak, as heteropteran mouthparts are properly called, these predators are able to inject a variety of enzymes that dissolve their prey into liquid. I felt like my hand was on fire!

The assassin bugs (family Reduviidae) are members of one of the largest families of true bugs. Most have raptorial front legs; others have hairy legs to help them hold on to prey. They range up to about a half-inch long. Some are generalists, but others are adapted to specific prey. For instance, an entire subfamily, the Ectrichodiinae, preys on millipedes. Nymphs of some species camouflage themselves with debris or the remains of their prey. The masked hunter (*Reduvius personatus*) covers itself with dust; if you find these in your home, know that bed bugs are among their preferred prey.

More to my liking are the stout little ambush bugs (family Reduviidae, subfamily Phymatinae). These look for all the world like miniature dinosaurs, and can be found on fall-blooming yellow flowers such as goldenrod. They are able to capture prey much larger than themselves, including bees, wasps, and flies, using their thickened raptorial forelegs.

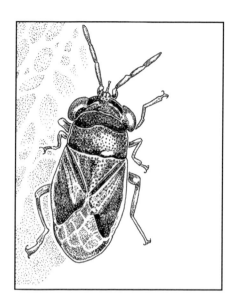

Big-eyed bug (*Geocoris* sp.)

Assassin bug (*Zelus* sp.)

Damsel bugs (family Nabidae) look like a smaller, more slender version of the assassin bug. Often shades of brown, they usually shelter in groundcovers and low-growing grasses. They are generalist predators of anything smaller than themselves, and are even known to cannibalize each other if no other food is available. As with so many other insects, adults overwinter in leaf litter.

Big-eyed bugs (family Geocoridae) are important garden predators of mites, flea beetles, whiteflies, aphids, small caterpillars, and insect eggs. They consume dozens of prey daily, and show great potential for biological control in agriculture. The adults are tiny, but true to their name, they have proportionally big eyes, so are fairly easy to recognize. If the eyes don't clue you in, they also walk with a distinctive "waggle" and emit a foul odor when handled. They are predatory, but survive on nectar and honeydew when there is a lack of prey. They are often found on flowers, and occasionally feed on plants.

First, the bad news: where they are abundant, the diminutive pirate bugs (family Anthocoridae) sometimes bite

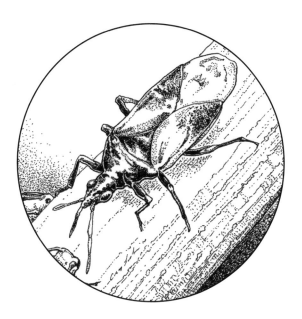

Minute pirate bug (*Anthocoris* sp.)

humans. Like mosquitoes, some people have an adverse reaction, others do not. Now the good news: These tiny bugs (under ¼ inch) are an important predator of corn earworm eggs. In addition, they prey on leafhoppers, thrips, and mites. Some prey on the larvae of leaf-mining flies and moths—the bane of my attempts at growing Swiss chard—so I generally protect my crop with a row cover, and the minute pirate bugs have to hunt elsewhere.

Accommodating a diversity of predatory bugs is easy. Many are found in association with small flowers in the aster and parsley families, which provide excellent places for assassin and ambush bugs to hunt. Damsel bugs frequent grasses, yet another reason to include these useful plants in the landscape. Minute pirate bugs are attracted to many kinds of flowers. But then again, aren't we all?

APPEARANCE Size varies. Many heteropterans have a prominent triangular "scutellum" visible on the dorsal side. Reduviidae, 0.2–1.6 in. (5–40 mm); Phymatinae, 0.2–0.5 in. (5–12 mm); Nabidae, 0.07–0.5 in. (3–12 mm); Geocoridae, 0.12–0.2 in. (3–5 mm); Anthocoridae, 0.06–0.5 in. (2–5 mm).

POPULAR NAMES True bugs, assassin and ambush bugs, damsel bugs, big-eyed bugs, minute pirate bugs.

SCIENTIFIC NAMES Order Hemiptera, suborders Heteroptera, families Reduviidae, Nabidae, Geocoridae, Anthocoridae.

COMMON TAXA The most easily spotted are the assassin bugs, which can be large, and often brightly colored. Slender brown nabids may also be seasonally abundant, but rarely noticed. Geocorids, while small, are easy to find and recognize. Anthocorids, while rarely noticed, are often abundant in gardens.

DISTRIBUTION Most families discussed here include common, widely distributed garden species: in America north of Mexico, 195 Reduviidae, of which 27 are in subfamily Phymatinae, 40 Nabidae, 30 Geocoridae, and 90 Anthocoridae.

GARDEN ACTIVITIES Many predatory bugs consume dozens of insect prey each day, including many common garden pests.

The Bugs Formerly Known as Homoptera

Aphids give live birth. Within a week, the new aphids are reproductive, and when the weather is hot, they reproduce quickly—up to five offspring each day for a month! It's a good thing that so many garden allies feed on sweet plump aphids.

The true bugs in this essay, all herbivores, are familiar to most gardeners. Unlike their relatives the heteropterans, both pairs of wings, when present, are fully membranous. For this reason their suborder was named Homoptera, "same-winged." (Taxonomic research has now separated them into two unrelated suborders. See notes at the essay's end.) Some of these can multiply without sexual reproduction (parthenogenesis); aphids often generate more aphids this way. Nymphs usually resemble adults; adults of many lack wings. Most are small, but they can become a big problem and often are not welcome in our gardens. Indeed, some are our most reviled insect pests, not only aphids but also scale insects, mealybugs, and whiteflies. But we have allied ourselves with homopterans in the past, and they have a role to play in our gardens, as we shall see.

Homopterans have been historically prominent in human affairs. The cochineal scale insect was the only good source of red coloring (carmine dye) for clothes and food until a synthetic alternative was invented in the late 19th

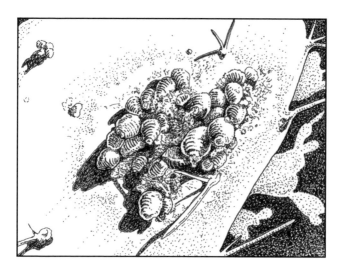

Cochineal scale (*Dactylopius coccus*) on cactus pad

century. Carmine dye was exported to Europe from the New World (South America), and was used to color fabric for royal clothing, religious vestments, and the British Army's red coats. It was historically used to color the Italian liqueur Campari. More recently, health concerns over the use of synthetic colorants have caused producers of certain foods and cosmetics to circle back to carmine dye. A related insect, the lac scale, forms masses that look like a dark, lumpy crust. This material is the source of the wood finish shellac. Did you know that before the advent of vinyl, record albums were pressed from shellac? The first plastic, Bakelite, was developed as a replacement for shellac products. Biblical scholars have suggested that the manna from heaven that sustained the hungry Israelites during their years in the desert was thought to be honeydew (sugary excreta), yes, from a scale insect.

The advent of modern biological control, as mentioned in the essay on lady beetles, was initiated by the accidental introduction of a homopteran, the cottony cushion scale from Australia. The introduction of two of its natural enemies, both native to the same region of Australia, the vedalia beetle and a parasitoid fly, have effectively controlled this pest of the California citrus groves since 1888.

This story shines a light on a basic truth here: these little bugs are food! Without them, many other insects and animals would go extinct, and the food web on which we and all life depends upon, would be severely disrupted. Most lady beetles, for instance, prey on aphids, scale, mealybugs, and other homopterans. Lacewings, parasitoid wasps, hover flies, and other insects that rely on these bugs as essential food are themselves critical food for others. Many fish, amphibians, and birds depend on insects, including the abundant former-homopterans, for all or part of their diet. I have watched hummingbirds gathering aphids when nestlings needed feeding, delighted that they were raising their young in my garden.

Remember, all homopterans are herbivorous—they eat plants. In response, plants have evolved numerous defenses, such as sticky hairs, waxy cuticles, and toxic compounds. Some plants even produce hormones that control homopteran growth. Herbivorous insects adapt to overcome certain plant defenses. The plants respond by increasing their defenses, and so on. The result is an "evolutionary arms race" in which each organism is adapting in response to the other, becoming more specialized as time goes by. The result is that many plant-feeding insects are highly host-specific.

How does this help you in your garden? Knowing that most aphid species are more or less host-specific allows for better control over where they can be tolerated. Well-established milkweed, for example, can survive the depredations of the specialized aphids that attack only it and a very few related plants. Therefore you could tuck in some milkweed plants, and the aphids it hosts will provide an alternate source of food for natural enemies (many are not picky about which species of aphid they prey upon) when populations are low on more valued plants, such as roses and broccoli. Annual German chamomile is another good choice as a "beneficial insect nursery" plant. Many plants in the mustard family are ideal, for example, wallflowers, sweet alyssum, and dame's rocket. Aging brassica plants in your vegetable patch can also serve as "nursery plants"; consider designating a couple for that purpose.

In many domesticated plants, defenses against herbivorous insects have been bred out (the unintended result of selecting for traits we like, such as tender, succulent foliage). Where we grow monocultures or even large patches, outbreaks of these bugs can result. Conversely, outbreaks may be prevented or mitigated if we "mix it up" in our gardens—that is, grow all sorts of plants and design our landscape so that mass plantings are avoided.

If these tiny herbivores appear on rosebuds, broccoli, or citrus, they clearly have to go. They can be difficult to kill with most pesticides, but there are other, better options. You can evict aphids with a stiff spray from a hose, or use insecticidal soap (follow the directions on the label). If you grow citrus and have trouble with scale insects, you may know that they can be smothered with horticultural oil. But alternatively, you could rein them in with natural predators by providing the nectar and pollen that these rely upon for some parts of their life cycle. Be sure to include ant control in your management plan; many ants "farm" aphids and other homopterans for their sweet honeydew and attack their "herd's" natural enemies.

I can count on large aphid colonies developing on the end-of-season sunflowers at the back of my garden. Instead of getting alarmed or reaching for a pesticide, I fetch my hand lens and investigate the community of allies that have gathered on the back of the fading leaves to feast on the pests. In my mild climate, I know that numerous species of beneficial insects are producing an overwintering population, ready to emerge early next spring.

If you spot what looks like white foam on your plants, you probably have spittle bugs in your garden. They show up on many plants, among them conifers, lavender, rosemary, beans, and strawberries. The foam, actually waste, has an acrid taste that discourages predators. It also makes the little nymphs within hard to spot by predators and protects them from temperature swings and dehydration. They are feeding on plant sap and do little actual damage. But if the foam bothers you, you can clean them off the plant with a blast from the hose. The adults are called froghoppers, which do little or no damage in gardens.

Cicadas are also homopterans and not, as some people think, a type of beetle. The nymphs live underground as root-feeding larvae, the larger species sometimes for many years (for example, the famous 17-year cicada). It takes so long for them to reach maturity because they feed on the watery fluid taken up by plant roots (xylem), a nutritionally poor food (rather than the sugary phloem most homopterans feed on). They may not be considered an ally by some gardeners, and can cause twig damage when laying eggs. But they are food for birds, and it isn't summer without the song of the cicada!

Cottony cushion scale (*Icerya purchasi*) with vedalia beetle (*Rodolia cardinalis*)

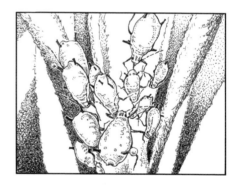

Milkweed aphid (*Aphis nerii*)

APPEARANCE Wings, when present, are membranous, and usually held rooflike over the body at rest. Size variable.

POPULAR NAMES Aphids, scale insects, mealybugs, whiteflies, cicadas. Hoppers: planthoppers, treehoppers, froghoppers.

SCIENTIFIC NAMES Formerly these were in a now obsolete suborder called Homoptera; researchers have now separated them into two unrelated suborders, Auchenorrhyncha and Sternorrhyncha (citing the lack of a common ancestor).

COMMON TAXA Auchenorrhyncha (cicadas, leafhoppers, treehoppers, plant-hoppers, spittlebugs); Sternorrhyncha (aphids, scale insects, psyllids, whiteflies, mealybugs, and phylloxera).

DISTRIBUTION Ubiquitous. Many are specialists, found on specific host plants.

GARDEN ACTIVITIES All are herbivorous. May feed on phloem, xylem, or sometimes pierce individual cells, depending on species.

The Odonata

Dragonfly nymphs have a prehensile, hinged lower "lip" tipped with claws that shoots out to capture prey. Adults can eat their weight in adult mosquitoes daily.

Within minutes of settling an old cast-iron tub into our perennial border and filling it with water, we were watching the aerial antics of several dragonflies as they swooped and darted about the garden. Intended to accommodate a few aquatic plants and provide water for wildlife, the new garden feature was clearly also going to provide plenty of easily viewed entertainment. We were delighted!

Dragonflies are large, heavy-bodied, strong fliers, with wings held flat at rest—the "cargo plane" of the order Odonata. Typically they are about 3 inches long. Closely related damselflies are more fragile-looking and about half as large. They are more accurately likened to helicopters. Slender, weak fliers, they generally hold their gossamer wings over the top of their body when at rest. In both groups, the aquatic larvae (naiads) can live a few weeks to several years overwintering in water. Adults, on the other hand, live only a few weeks (in temperate species), and die in winter.

These fascinating creatures are an ancient lineage of insects, not much changed since the time of dinosaurs. The main difference is size; fossil dragonflies with wing spans of over two feet have been found! Although they have powerful, serrated jaws ("mandibles"), dragonflies do not bite people unless

Flame skimmer dragonfly (*Libellula saturata*)

roughly handled. With their large compound eyes, they are efficient hunt-ers, both as naiads and adults. If you are lucky, you may one day witness a naiad's exoskeleton splitting down the back and the magical emergence of a brand-new dragonfly, as its crumpled wings unfold in slow motion.

Adults are voracious predators, and will eat just about any flying insect. Flight patterns are usually related to hunting style. Most catch their prey on the wing in a basket formed with their legs. Some dragonflies exhibit "hawking" behavior, relentlessly pursuing prey, while others "sally," darting out from vegetation, then returning to their perch. Others are "hover-gleaners," picking insects from vege-tation while in flight. Naiads are also predators and often hunt by ambush. Those in my tub are camouflaged with algae, allowing them to sneak up on their prey, when their prehensile lip (labium) shoots forward to snag the hapless victim.

While I live fairly close to a large wetland and have undoubtedly lured some of its residents to my yard, many dragonflies will travel a long distance from permanent water to reach a garden water feature. The more delicate damselflies are harder to attract if you are far from water. Garden water features can also attract all sorts of other aquatic insects. I have observed back-swimmers, creeping water bugs, and water striders in mine. Kneeling next to my backyard tub, I always have an amusing circus to observe.

If you want to install a garden water feature, here are a few tips. Create one that varies in depth; if you cannot put in a side shelf, just elevate some potted water-loving plants (on overturned pots, rocks, bricks, or cinder blocks). Include a variety of water plants, avoiding those that are invasive in your region. Emergent vegetation such as rushes and sedges provide convenient perching places for adults as well as a safe spot for naiads to molt. A few flat rocks near the edge offer basking spots, not to mention access to the water for bees and other insects and wildlife. Just remember that if you want frogs, you should empty your garden pond annually in fall to discourage hungry dragonfly naiads from decimating the tadpole population.

One concern you will have is mosquitoes, which lay individual or tiny "rafts" of eggs in the water, which soon hatch larvae. Odonata naiads dine on the larvae, but may not be able to exert sufficient control. One solution is the addition of a splashing fountain, which disturbs the water's surface enough to discourage mosquitoes from depositing their eggs but allows other aquatic insects such as dragonflies to lay theirs.

Another solution, which works for me, is to regularly add *Bti* (*Bacillus thuringiensis* subspecies *israelensis*) to my pond. This bacteria kills only mosquito larvae. It comes in "donuts" or granules, with instructions on when and how to add it to your water garden; follow these carefully.

Pacific forktail damselfly (*Ischnura cervula*)

I am always thrilled to see wheeling dragonflies dip down to lay their eggs, and later, the appearance of tiny naiads. It is comical to watch males skirmish in their efforts to access a female, often butting each other aside as they jockey for position. And if you have damselflies, look closely. The curious circle formed by mating damselflies is said to be where the romantic heart symbol comes from.

Every region has its most common dragonflies, but whichever species come to your garden are sure to please. In my garden, the cardinal meadowhawk is a frequent visitor. I can approach very closely before it flits away, only to return minutes later to the same reed. The large red dragonfly patrolling overhead is a flame skimmer. Pond damsels are occasional visitors, and include the sapphire-colored vivid dancer and familiar bluet. I sometimes see pacific forktails, blue-eyed darners, and other species I have yet to identify. The shimmering flight of dragonflies provides hours of entertainment, and I can no longer imagine my garden without these welcome guests any more than I can imagine my home without friends!

APPEARANCE Aquatic larvae, known as naiads. Adult dragonflies generally stout-bodied; wings held flat at rest; large compound eyes often meet in the middle. Damselflies usually slender; wings generally folded over body; eyes set widely apart.

POPULAR NAMES Dragonflies include skimmers, emeralds and baskettails, cruisers, spiketails, clubtails, darners, and petaltails. Damselflies include broad-winged damsels, pond damsels, and spreadwings.

SCIENTIFIC NAMES Order Odonata, suborders Anisoptera (dragonflies) and Zygoptera (damselflies).

COMMON TAXA Dragonflies: flame skimmer (*Libellula saturata*), cardinal meadowhawk (*Sympetrum illotum*), blue-eyed darner (*Aeshna multicolor*). Damselflies: vivid dancer (*Argia vivida*), Pacific forktail (*Ischnura cervula*).

DISTRIBUTION Over 5400 species worldwide; many tropical species. About 435 species in America north of Mexico. There are more species in southern regions; however, some species are only found in northern climes.

GARDEN ACTIVITIES Both adults and naiads are carnivorous, eating a lot of other insects, including larval and adult mosquitoes.

CRICKETS, GRASSHOPPERS & KATYDIDS

The Merry Musicians

Grasshoppers are nutritious food for many types of predators and they have evolved numerous defenses to avoid being eaten, including camouflage and acrid secretions. If those aren't effective, their underwings are often brightly colored, startling predators just long enough for the grasshoppers to escape.

A few years ago, I was the recipient of a lovely illustrated children's book, *A Cricket in the Grass and Other Stories,* authored by a gardening friend, Philip Van Soelen. The cricket in the title makes an appearance only at the beginning of the wordless book, as a passing toad promptly eats it! The toad soon meets a similar fate. And so the chain continues, a lesson in the importance of all creatures—in field, forest, or garden—in the food web.

Orthoptera—that is, the grasshoppers, katydids, and crickets—almost always fall into the category of prey. While a few species may cause damage, sometimes extensive (especially in agriculture), most are fairly innocuous in gardens. They serve as an important food source for many species of birds, mammals, amphibians, and reptiles, in the wild as well as in our gardens. They are nutritious food (high in protein and low in cholesterol) for people in numerous cultures around the world. Consuming insects, called "entomophagy," is becoming increasingly contemplated as we seek new ways to feed a hungry planet.

Snowy tree cricket—female above, male below (*Oecanthus fultoni*)

All orthoptera possess hind legs adapted for leaping ("saltatorial" legs), a key characteristic for the order, although they are not all jumpers. They may be winged or wingless; if they have wings, these are often leathery. All have chewing mouthparts and most feed on plants. Some are omnivorous, and a few feed on decomposing matter. Typically their lifespan is a year. They are found in many habitats, but are most numerous in arid climates.

This is an order of singers! To attract females, males chirp, trill, and click by rubbing together their forewings, or hind legs and forewings, depending on the subfamily to which they belong. The loudest males are also often the largest, and the most successful in attracting mates.

With a culture fascinated by many things entomological, Japan has a long history of appreciating orthopteran music. Elaborate cages are built to house crickets, and although small, they can fetch large sums of money. At summer's end, at an "Insect Hearing Festival," captive crickets are released into nature, resulting in a joyous cacophony. In other cultures, crickets are thought to be harbingers of good luck. An entire 2000-year period of Chinese history is known as Cricket Culture. Among American Indian tribes, crickets and grasshoppers play different roles. The Cherokee, for instance, view crickets as animals to be emulated. An Algonquian legend about how grasshoppers were created suggests that the acrid juice they spit out is from chewing tobacco, an important herb in that culture.

The large Acrididae family of grasshoppers is most abundant in arid, grassy habitats, and feeds on grasses and non-woody, herbaceous plants (forbs). Most adults are winged and strong flyers, with short antennae. The spine-breasted locusts include the aptly named devastating grasshopper, which periodically wreaks havoc on agricultural crops, leading to terrible famines. Like other true locusts, it gathers in large groups or swarms and exhibits migratory behaviors that include physical changes. Most grasshoppers, however, have little economic impact. While over 550 species are found in North America, very few regularly cause significant damage to crops or gardens.

Band-winged grasshoppers are effectively camouflaged to blend into the background of the arid places they favor: deserts, canyons, arid grasslands, and drought-tolerant gardens in dry climates. But in flight they display beautifully colored hind wings, a secondary defense intended to startle predators. Slant-faced grasshoppers are easily recognized by their characteristic pointed heads and slanted faces, and include several desert species. The bright colors of the lubber grasshoppers (family Romaleidae), heavy-bodied insects with stumpy wings, advertise their bitter taste with "aposematic" coloring (as seen in lady beetles and others), and are an exception to the rule that orthopterans taste good.

The katydids (family Tettigoniidae) are beautiful insects, always welcome in my garden. They have very long antennae. The shield-back katydid, which superficially resembles some cricket species, can be distinguished from crickets by its long antennae. Katydids can be hard to spot, as many species mimic leaves (even having "veins") and they are normally

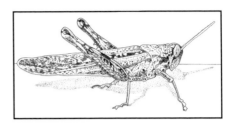

Grey bird grasshopper (*Schistocerca nitens*)

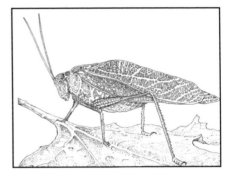

California angle-winged katydid
(*Microcentrum californicum*)

active after darkness falls. By rubbing the base of their forewings together, males are able to produce a variety of notes. In contrast to most of their orthopteran relatives, female katydids are also able to produce sounds in response. Most katydids are generalists (some eat only leaves, others also eat insect eggs, aphids, and small bugs). They are not usually present in sufficient numbers to be considered destructive. They are just beautiful members of the ecosystem.

Cave and camel crickets (family Rhaphidophoridae) are ground-dwellers, although a few species are found in trees. Their colors are subdued. Recognize them by their especially long antennae and curious "humped" backs. Wingless, they are a silent family. They prefer a moist environment and their primary diet is fungi, although as omnivores, they may eat insects, fabric, and other items. They show up in woodland gardens, and also frequent damp basements and garages.

As for what we consider the "true crickets," these are common ground-dwelling field and tree crickets (family Gryllidae) found throughout North America. They are the most accomplished of musical artists. Found in fields and grasslands, in addition to gardens, they sometimes have years of great abundance and can cause significant crop losses as a result.

Despite their name, the frequently heard snowy tree crickets are not white but rather pale jade-green. Males have a soft trill that serenades us from trees and shrubs in the evening hours, a sound so soothing and iconic that it is their song that is most often recorded for films and videos. Another name for them is "thermometer crickets." Simply count the number of chirps in 14 seconds, then add 40 to get the temperature in Fahrenheit. If you need another reason to welcome them to your garden, although they are mainly herbivorous, they also eat aphids! One of the great pleasures of a summer evening is to go out with a young guest and a flashlight and track down some snowy tree crickets.

Crickets and their kin are primarily herbivores, and females of some can damage twigs when they lay their eggs. On the other hand, these often-beautiful creatures offer some benefits to our gardens. And some are omnivores, creating the potential for your garden to become a setting for "dinner and a concert."

APPEARANCE All orthopterans have jumping hind legs. Acrididae, 0.35–3 in. (9–80 mm), most 0.6–1.2 in. (15–30 mm), some may reach up to 2 in. (50 cm). Crickets are usually dorso-ventrally flattened; males are distinctly musical.

POPULAR NAMES Grasshoppers, locusts, katydids, crickets.

SCIENTIFIC NAMES Order Orthoptera; two divisions: Caelifera (short-horned grasshoppers and allies) and Ensifera (crickets, katydids, and allies). Ensifera are divided into infraorder Gryllidea ("true crickets") and infraorder Tettigoniidea, (katydids, camel crickets, and relatives).

COMMON TAXA Short-horned grasshoppers (family Acrididae), subfamilies include band-winged grasshoppers (Oedipodinae), slant-faced grasshoppers (Gompho-cerinae), and migratory bird locusts (Cyrtacanthacridinae), and two others. Family Gryllidae includes the snowy tree crickets and field crickets. Infraorder Tettigoniidea includes katydids, raspy crickets, and Jerusalem crickets. Camel crickets are family Rhaphidophoridae.

DISTRIBUTION Caelifera includes 660 species in 5 families in North America, 620 of these are in Acrididae. Ensifera includes 1200 species in 10 families in North America.

GARDEN ACTIVITIES Insects in Caelifera feed almost exclusively on plants; Ensifera includes the katydids (Tettigoniidae), which may be omnivorous.

Praying ... or is it Preying? Mantises

Hundreds of nymphs can emerge from a single foamy mantis egg mass (ootheca), running in all directions. Those that don't disperse quickly enough may be eaten by their hungry siblings. Most baby mantises feed on small insects like aphids and fruit flies.

Late on a balmy summer evening I saw a pair of mating mantises hunting night-flying insects near the porch light. Approaching to take a closer look, I was astonished to discover a hapless, headless male closely clasping his paramour even in death. The female had bitten off his head while mating. Although this gruesome behavior, known as sexual cannibalism, is well-documented among mantises in captivity, it is actually rare in nature where food is more abundant.

Mantises, it seems, are always hungry. Tiny young nymphs the size of an ant emerge from a foamy proteinaceous egg mass (an ootheca) that looks like beige Styrofoam, and almost immediately scatter widely. In a confined space, they will turn on each other for a convenient meal. Otherwise, carnivorous young mantises dine on tiny prey such as fruit flies and aphids, before switching to larger insects as they grow. In the garden, mantises have a reputation for eating pests, however as generalist predators they eat whatever meal wanders by. Like most mantises, garden species are ambush hunters and can often be found hanging around flowers, a rich source of insect prey such as bees, flies, and butterflies.

Mantises have extremely good vision, thanks to large, wide-set eyes and a head that can rotate almost 180 degrees, set on an elongated thorax. Two

pinpoint "pupils" (aptly called pseudopupils) give them a unique trait—they seem to meet your gaze. While their reliance on vision for hunting means that they hunt during daylight hours, it is rare to see a mantis in flight during the day. Males seeking mates tend to fly at night when they are protected from birds and other predators and, like the pair that I spotted, they are sometimes drawn to lights.

These amazing creatures possess powerful front legs armed with spines. From a perfectly still position, they move their raptorial legs at lightning speed to capture and hold passing insects. This position has led to the common name of "praying" mantis, more properly reserved for a single species,

Ground mantis (*Litaneutria minor*)

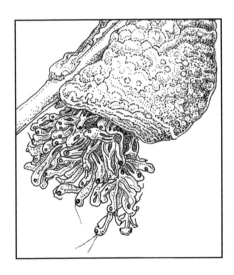

Chinese mantis (*Tenodera sinensis*) nymphs emerging from ootheca

the European *Mantis religiosa*. Stories of tropical mantises capturing other animals are often exaggerated, but large mantises do occasionally attack lizards, frogs, and rodents, as well as birds. If the prey is large and struggling, mantises begin by eating the victim's head first, then leisurely completing their meal.

Long thought to be related to grasshoppers, katydids, and crickets, today mantises are known to be more closely allied to cockroaches. Perhaps this is actually not surprising, given that both mantises and cockroaches produce their eggs in the mass (ootheca) mentioned above. Like cockroaches and grasshoppers, mantises are hemimetabolous. Juveniles resemble wingless adults, developing longer wings with each succeeding stage until adulthood. In some species, adult female mantises are flightless and develop engorged abdomens as they approach the time to produce the ootheca. Males are much smaller and are more likely to fly. Mantises live between six months to a year, although females die soon after producing the ootheca.

Several garden mantis species are introduced, notably the European mantis, the Mediterranean mantis, and the Chinese mantis. It appears that established populations of Mediterranean mantises may be competing with native mantises in southern California, yet another example of the risk of introducing species where they do not belong. Egg cases available for sale from garden suppliers and nurseries are typically those of the Chinese mantises. If you are patient, however, native mantises eventually show up in your habitat garden, if they belong there. While they are generalist predators, they provide plenty of amusement even they are not controlling pests.

In most of the United States, the native Carolina mantis is a common garden resident. These range from leafy green to twig brown, which provides some measure of camouflage. In the western United States, small desert species can be spotted running about rapidly on the ground in what seems to be a most un-mantislike fashion. The only mantis native to Canada is a ground mantis, which turns out to range as far south as Mexico.

The diversity of mantis species is greatest in the tropics. Tropical mantises take all sorts of bizarre forms. Some resemble lichens, flowers, or leaves, providing great examples of masquerade, a type of camouflage in which an animal resembles an inanimate object in the environment. This provides the perfect disguise when hunting, while simultaneously protecting themselves from predators.

Mantises are themselves preyed upon by other garden visitors, particularly birds. If disturbed or threatened, they defend themselves by unfurling their wings; dark markings on their underwings appear to be the eyes of a larger animal and are meant to startle predators.

The jury is out on whether mantises eat more pests than allies, or perhaps have no real impact on the garden food web. Certainly they are an intriguing example of garden fauna; children, especially, are fascinated by them. With their rotating heads and alert-looking eyes, they are a popular insect "pet" and can even become accustomed to being handled. When I was a child, my mother generously allowed me to keep pet mantises on our kitchen windowsill, where they caught any unaware flies buzzing about, and contributed to a lifelong love of all things entomological.

APPEARANCE Sizes vary, 0.4–6.7 in. (10–170 mm). Green Mediterranean mantises are up to 2.5 in. (60 mm). European mantises are 2–2.4 in. (50–60 mm) and exhibit great variation in color—green, brown, or yellowish. The Chinese mantis may be up to 4 in. (100 mm) in length, often brown. Carolina mantises are brown, gray, or green; about 1.8–2.4 in. (45–60 mm); wings of females are truncated. The agile ground mantis, well camouflaged gray or brown, is less than 1.4 in. (35 mm).

POPULAR NAMES Praying or preying mantis.

SCIENTIFIC NAMES Order Mantodea, now split into many families.

COMMON TAXA Mediterranean mantis (*Iris oratoria*), European mantis (*Mantis religiosa*), and Chinese mantis (*Tenodera sinensis*) are all introduced. Natives include the Carolina mantis (*Stagmomantis carolina*) and the agile ground mantis (*Litaneutria minor*).

DISTRIBUTION America north of Mexico, 20 species of Mantodea in 7 families; about 2300 species worldwide; most are tropical.

GARDEN ACTIVITIES They eat whatever they can catch. Garden mantises are often found perched on flowers, ambushing any unaware visitors.

Delicate Beauty, Ferocious Killers

Beautiful green lacewings lay their eggs at the tips of delicate threadlike "stalks"—an adaptation that hinders predation and parasitism of their eggs by natural enemies looking for a quick meal.

When I was a small child, my father one day plucked a leaf from our backyard apple tree and flipped it over to show me a few trembling inch-long threads tipped with delicate pearls. These amazing, frail jewels were the improbable-looking eggs of green lacewings.

Green and brown lacewings belong to the order Neuroptera, along with some intriguingly named creatures such as dustywings, antlions, owlflies, and mantisflies. Snakeflies are related, as are dobsonflies, fishflies, and alderflies. The most common in gardens are lacewings, dustywings, and snakeflies.

Green lacewing adults are familiar to most gardeners. They are quite small, approximately ⅜ of an inch long, and their diaphanous mint-green wings blend into foliage. That alone makes them hard to spot, plus they often like to hang around trees, above our normal field of vision. Nocturnal by nature, they are sometimes attracted to porch lights. Occasionally, though, you may see adult green lacewings visiting your garden flowers during daylight hours. An observant gardener may also spot the smaller, more reliably nocturnal brown lacewings, and the curiously long-necked snakeflies. As for

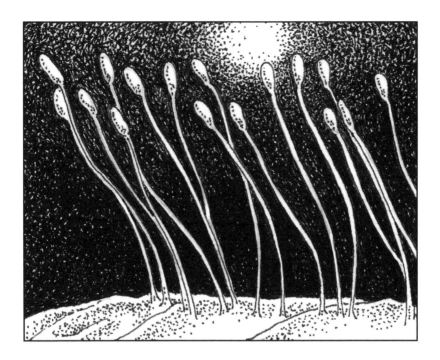

Green lacewing (*Chrysoperla carnea*) eggs

the minute dustywings, if we notice them at all, they are usually mistaken for whiteflies.

The life cycles of these insects are influenced by temperature. It is not unusual for neuropteran adults to live for four to six weeks in good weather. Some species may produce several generations per year in mild climates. Some overwinter as adults, often in leaf litter. Most are predatory as larvae, and many also as adults. Some feed exclusively on pollen, nectar, and honeydew as adults. But garden neuropterans are primarily consumers of other insects, preying on all sorts of arthropods, including aphids, scale insects, whitefly, and mealybugs. They may also dine on spider mites, insect eggs, beetle larvae, thrips, and small caterpillars. Some species are known to detect the larvae of leaf miners, piercing the leaves to reach their prey.

The larvae of green lacewings (family Chrysopidae) really do resemble small alligators, although a common name for them is "aphid lion." Look for them among their somewhat sedentary prey (often aphids). Their mandibles are equipped to pierce prey, injecting paralyzing, liquefying venom that enables the larva to suck out the victim's fluids. A favorite plant for egg-laying is borage, an easy plant to incorporate into a vegetable patch, herb garden, or flower bed. As a bonus, bees love borage.

Some species of green lacewing larvae have fascinating camouflage. They disguise themselves with debris such as bits of plant material and the empty hulls of their prey. Occasionally, a disguised "wolf in sheep's clothing" will catch my eye—a well-fed lacewing larvae among a colony of mealybugs, looking exactly like a mealybug, until one notices its speed and developed legs. It is thought by some scientists that the larvae adopt disguises not to make themselves invisible to their prey, who seem oblivious to the presence of the "wolf" in their midst, but to hide from their own predators. In any event, the voracious larvae can eat hundreds of pests during their development.

Although smaller and drab-colored, brown lacewings (family Hemerobiidae) are readily recognized by alert gardeners. Unlike the green lacewings, they lay their eggs singly on tree leaves or bark. The larvae are very similar to green lacewings', but with less prominent mandibles. Brown lacewings are important garden insects, as both adults and larvae are predators, primarily of scale insects, aphids, and whitefly nymphs. Many species are restricted to certain plants, suggesting that they may be specialist predators. They are long-lived, have a high reproductive capacity, and voracious appetites.

Dustywings (family Coniopterygidae) are easily distinguished from whiteflies by their wing position at rest, exhibiting the typical "roof" shape of their relatives the lacewings (whiteflies, in contrast, hold their wings much flatter).

Green lacewing larva

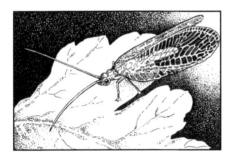

Green lacewing adult

This tiny predator, most active at dawn and dusk, is a tree-dweller. It has an appetite for spider mites, as well as scale insects and arthropod eggs. The tiny larvae resemble minute lacewing larvae. More common than once thought, they are found primarily in warmer parts of the Pacific Northwest.

Snakeflies (family Raphidiidae) invariably provoke anyone who sees them for the first time to recoil. They have elongated "necks" and heads that can turn much like a praying mantises'; females have a long ovipositor that looks like a stinger, although it is harmless. The peculiar larvae, sometimes found under bark, have the unique habit of running backward when startled. Adults can be spotted visiting flowers, but are more often found in trees and shrubs, where they are generalist predators as both larvae and adults.

You can assume that many of these allies are present in your garden. They can inhabit many different niches, so keep an eye out. If you do not spot any, green lacewings are one of the few insects worth buying and introducing to your garden, where they will establish permanent

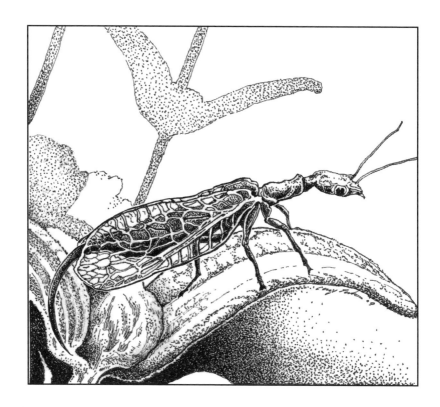

populations. A layered environment of trees and shrubs, annuals and perennials, provides them with the habitat they need.

Some years ago, a friend suggested that anyone obsessed with insects (did she mean me, I wondered?) would be well-advised to acquire a pair of the close-focusing binoculars popular with butterfly enthusiasts. Peering at the lacewings on my flowers, with their delicate shimmering wings and great golden eyes, I am glad to have taken her advice.

APPEARANCE Varies; all have characteristic net-veined wings as adults. Larvae of lacewings look like miniature alligators. Snakefly larvae have large heads, and projecting mandibles; they run backward when startled. Green lacewings, 0.5–0.75 in. (12–20 mm); brown lacewings, 0.4–0.6 in. (10–15 mm); snakeflies, 0.5–1 in. (12–25 mm); dustywings, 0.07–0.2 in. (2–5 mm).

POPULAR NAMES Lacewings, dustywings, snakeflies.

SCIENTIFIC NAMES Order Neuroptera includes green and brown lacewings, mantis-flies, antlions, owlflies, and dustywings. Superorder Neuropterida includes two additional orders: Raphidioptera (snakeflies) and Megaloptera (fishflies, dobsonflies, and alderflies).

COMMON TAXA Green lacewing (*Chrysoperla* spp.), brown lacewing (*Hemerobius* spp.), snakefly (*Agulla* spp.).

DISTRIBUTION Worldwide, about 6000 species.

GARDEN ACTIVITIES Many common garden species prey on a broad range of soft-bodied insects, especially aphids and other small arthropods. Some adults do not feed, or feed only on nectar.

Ants . . . As Allies?

Ant colonies have a complex system of communication mediated by pheromones (chemicals) they excrete. Instructions for trail construction, location of food sources, alarms, and recognition of colony mates are all controlled by pheromones.

In Aesop's fable, the ant is held up as a paragon of industry. Carpenters, weavers, gardeners, harvesters, drivers, and honeypot ants all uphold the reputation of the hard-working cooperative society. The ant realm is endlessly fascinating, as the plethora of books and nature programs on the subject attests. When ants invade our homes, however, our admiration ceases and we quickly perceive them to be pests. In a garden setting, however, most ant species are beneficial, crucial to aerating and nutrient cycling in the soil and the overall health of the ecosystem. There are probably tens of thousands of ants in a typical home garden.

Pest ants in gardens are usually those feeding on honeydew, produced by homopterans such as scale insects, aphids, and mealybugs. On citrus trees and other garden plants, that confluence is a problem, because the ants protect homopterans from natural enemies. All that sugary honeydew is also the perfect place for black sooty mold to develop, which blocks the sunlight the leaves need to photosynthesize. But in many cases, the role of ants in the garden is not so easily determined. For instance, some species that are principally beneficial sometimes supplement their diet with honeydew, albeit only when other resources are scarce. And while you certainly do not want carpenter ants chewing on your house, garage, or garden shed, they may provide a service in the landscape, where they help break down dead wood. Encourage them to "know

Velvety ant (*Liometopum occidentale*)

their place" by never leaning or stacking wood or firewood adjacent to your home or outbuildings.

Ants (family Formicidae) are far more diverse that most of us realize. Although some species are widespread, others occur only in a limited region. Many ants require specific conditions; species found in cool, damp areas of the Pacific Northwest are different than those found in the Southwest. Some common species are rarely noticed. Honeypot ants, for instance, are found in many arid and semi-arid areas, where they are fairly common. However, the unusual workers, their abdomens plump with nectar, are rarely seen because they are ensconced underground. Nor are all ants native; Argentine ants are a problem wherever they are found in the United States, as they aggressively attack and kill any native ants, in addition to protecting homopteran colonies from attack. They are also a frequent invader of our homes.

One observer described an ant colony as "a factory inside a fortress"—an accurate portrayal. Ants are highly evolved social insects. Duties are divided among castes of adults, which typically include reproductive queens, sterile female workers (sometimes of several different castes, each with its assigned task), and short-lived males (whose only role is reproduction). The nuptial flight is the evocative name for the mating that occurs when the winged females and males emerge from the nest. Unlike honey bees, some ant species have multiple queens (the case with Argentine ants), which is why it can be difficult to eliminate them. Individual ants can live up to several years. Queens are especially long-lived, in excess of 30 years in some carpenter ants!

Ecologically, ants are an essential taxa of insects. In addition to their key role in the aeration of soil and nutrient cycling, they are known to sometimes act in pollination, distribution of seeds, and pest regulation. Many are scavengers. Depending on species, they may also feed on plants, invertebrates, honeydew, fungi, or a combination of these.

Honeypot ant (*Myrmecocystus* sp.)

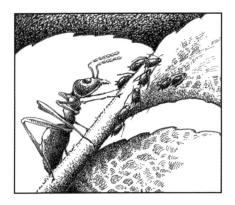

Argentine ant (*Linepithema humile*)

Some plants have evolved to attract ants, with extrafloral nectaries (EFNs), or special packets of nutrients, called food bodies (on plants) or elaiosomes (on seeds). The ants then aggressively attack any approaching herbivores. Some examples are peonies, fruit trees such as peaches, plums, and cherries, and, in the vegetable garden, squash, melons, and gourds.

It is well-known that some ant species defend aphids and other homopterans from their natural enemies in order to feed on the honeydew they produce. Yet some of the most useful ants, from a biological control perspective, are species that only turn to honeydew when other food sources are scarce. As with so many other insects, observation is key. Ants in your citrus will need to be controlled, but native ants in other areas of your garden may well be left alone.

To evict unwanted ants, you may have to use a combination of methods. For maximum effect, concentrate your control efforts in the late winter and early spring when the colonies are less active. If possible, reduce or eliminate their food source. Finding and eliminating nests, obviously, will have a more lasting effect. Avoid chemical warfare, which can cause unintended harm to other creatures, the soil food web, and desired plants. Borax baits are widely available. Boiling water poured over an ant mound or diatomaceous earth (an abrasive powder) sprinkled around ant mounds and trails, and around vulnerable plants, can be effective.

I try to follow the advice, perhaps apocryphal, of renowned biologist E. O. Wilson, who has devoted his life to the study of ants. When asked his opinion on what to do about ants in the house, he is reputed to have answered, "provide some crumbs and get out a magnifying glass!"

APPEARANCE When you notice ants moving their brood, you are seeing larvae and pupae, not the microscopic eggs. Adult ants are under 1 in. (25 mm) long. Ants may be black, brown, red, yellow, or combinations of these colors.

POPULAR NAMES Ants.

SCIENTIFIC NAMES Ants are all in a single family, Formicidae, in the order Hymenoptera.

COMMON TAXA Carpenter ants (*Camponotus* spp.), acrobat ants (*Crematogaster* spp.), field ants (*Formica* spp.). *Lasius*, *Aphaenogaster*, and *Tapinoma* are also among the taxa widely distributed in North America. Very few species of ants are pests. Introduced pest species include the imported fire ant (*Solenopsis invicta*) and the Argentine ant (*Linepithema humile*).

DISTRIBUTION Almost 1500 species in North America; over 15,000 species worldwide.

GARDEN ACTIVITIES Most are generalists that subsist on honeydew, nectar, and scavenged protein; a few groups are specialists, such as fungus-growers or army ants that prey on other ants.

6

THE GROUND CREW AND BEYOND

Centipedes, Spiders, Other Arthropods, Galls & Pathogens

We have met many insects in these pages, but quite a few of the invertebrate animals we encounter and think of as bugs are not insects at all. Insects, as we have learned, have three pairs of legs, while many of their arthropod (the large Arthropoda phylum, "joint-footed") relatives have more, sometimes far more! This chapter will delve into some of these other participants in the web of life in our gardens.

In case you are a bit fuzzy on high school biology, the invertebrates are distinguished by having no backbone. Arthropods, contained therein, have several notable features: exoskeletons of chitin, segmented bodies, and jointed limbs. There five different groups (subphyla) of arthropods. So far, we have spent the most time with Hexapoda, which includes the insects and few other rarely noticed six-footed groups (although we met the springtails in Chapter 1). We also covered Crustacea's roly-polys and a few aquatic forms. One subphylum, the trilobites, Trilobita, is extinct. Let's now meet the Myriapoda and Chelicerata.

First up are the aptly named Myriapoda (many-footed), which includes the centipedes, millipedes, and symphylans. They are frequently present in our gardens and soils. Once you understand them better, you will consider them part of your yard's support staff.

We will explore Chelicerata selectively, for only some are commonly found in our gardens—principally spiders and mites. And although some people may find them frightening, spiders are an integral component of a healthy garden food web and a critical element of successful biological control strategies. Some can survive months without food, but when they eat, they are all carnivores.

One spider relative deserves special mention: the harvestman or "daddy long-legs" (family Opiliones). Unlike the "daddy long-legs *spiders*"—web builders that mostly hang out in cellars and stairways—these outdoor-dwellers do not make silk or spin webs. They are found elegantly striding around your garden catching a variety of small insects as well as consuming insect waste and carcasses. They have extremely long, slender legs and, unlike spiders, only one apparent body part. They are neither venemous nor harmful to humans.

I have also provided an essay on the fascinating galls, which are very often induced by invertebrates (generally wasps or flies). Smaller invertebrates that lack legs altogether are also found in home landscapes, notably the ubiquitous slugs and snails (in their own phylum, Mollusca). By now, you know that I am going to say that they are an important part of the food web, but I have a few other insights for you. Last but not least, I have included an essay on pathogens that attack invertebrates, as they also figure in a garden's ecosystem.

The Many-Footed

Look closely—but don't touch! Each succeeding pair of a centipede's legs, from front to back, is longer than the pair before it. This helps them to not trip over their own legs when they run.

Myriapods have, as you might expect, myriad feet on flexible segmented bodies. Lacking the protective calcium-enhanced exoskeleton of the crustaceans or the waxy cuticle of insects, they are only found where there is moisture, such as in leaf litter, under rocks, dead logs, and other damp spots. All reproduce via eggs; males produce a spermatophore, a packet of sperm, delivered to the female externally, which she retains until ready to lay eggs. Juveniles have only a few segments and add more with each molt. Myriapods tend to be longer-lived than most the other arthropods we have talked about thus far. Millipedes may live up to ten years, while centipedes live one to six years.

While the "pair-footed" millipedes, class Diplopoda, and the "lip-footed" centipedes (more on that name in a moment), class Chilopoda, look superficially similar, they are very different animals. The smallest myriapods are the symphylans, class Symphyla; they're called "garden centipedes."

Most millipedes have long cylindrical bodies, and might be mistaken for worms, except for their feet. Although their name means "thousand-footed," they have far fewer feet than that. At first glance, it seems they have two pairs of legs per segment. But the segments are fused in pairs with one pair of legs

Centipede (*Theatops californiensis*)

each. Most millipedes support 20 or more pairs of segments. Do the math—that's 80 individual legs!

A few species are herbivorous, and can be problematic in gardens, greenhouses, and agricultural systems. However, the majority of millipedes are harmless detritivores, consuming decaying organic matter, particularly helping to break down leaf litter. In this respect, they do similar work to the earthworms we gardeners so value.

The American giant millipede (*Narceus americanus*) is gray and large, up to 4 inches long. It is widely distributed throughout North America. In my garden, I am more likely to spot a *Tylobolus* millipede. Reddish brown, cylindrical, and generally around 3 inches long, seven similar species are only found (with one exception) from where I live in northern California up into the Pacific Northwest. Slow-moving, a millipede's primary defense from predators is to coil up, but many also excrete protective toxins as a secondary defense.

Centipedes are one of the largest invertebrate predators. Some tropical species can even reach over a foot in length! Thankfully, most garden species are far smaller, usually in the 1- to 3-inch range. They are fairly easy to distinguish from millipedes, as their bodies are generally flattened top-to-bottom (dorsoventrally) and their antennae are long. As hunting predators, they can run fast. While centipedes only have one pair of legs per segment, they are usually moving so quickly it would be difficult to count! Centipedes—the "hundred-footed"—have between 15 and 177 pairs of legs, always an odd number. In other words, there are no centipedes with literally 100 feet. The front pair of legs is modified into venom-injecting claws, or "lip-feet" (called forcipules), that are used as accessory mouthparts to capture, paralyze, or kill prey.

Centipedes are carnivores in garden environments, preying on many insects, as well as spiders and other centipedes. The clawlike pincers can inject venom into fingers of the curious and the bare feet of the unwary. Most have subdued coloration—but not *Theatops californiensis*. These are bright orange-red,

Millipede (*Tylobolus* sp.)

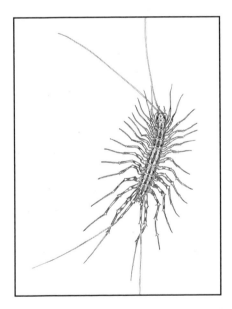

House centipede (*Scutigera coleoptrata*)

fortunately warning the gardener from picking them up! In North America, most centipedes don't inflect pain any worse than a wasp sting (some may not even penetrate human skin), barring allergies.

The scary-looking introduced house centipedes (*Scutigera coleoptrata*) are harmless, at least to humans. Like other myriapods, they are found in damp spots. (In our houses, they favor shower and bathtub drains and pipes.) Outdoors or indoors, resist the temptation to kill them. They are carnivores, eating all sorts of insects.

Garden centipedes (class Symphyla) live in the soil. Usually less than a third of an inch length, they have 12 pairs of legs, and move rapidly. They are unpigmented and translucent. Most eat decaying organic matter. Unfortunately, many are omnivores and can be root-nibblers. They can cause significant damage in a garden or agricultural setting and, if they get established, are difficult to control. It is best not to encourage them. I play it safe by adding compost only on top of my soil rather than digging in partially decomposed material that might attract them. Then I let the earthworms and other soil inhabitants do the digging.

APPEARANCE The many-footed are long, with multiple pairs of legs. Often shades of gray, black, or brown; some centipedes are bright red.

POPULAR NAMES Centipedes, millipedes, garden centipedes.

SCIENTIFIC NAMES Phylum Arthropoda, subphylum Myriapoda, classes Chilopoda (centipedes), Diplopoda (millipedes), Symphyla (garden centipedes).

COMMON TAXA House centipede (*Scutigera coleoptrata*), American giant millipede (*Narceus americanus*).

DISTRIBUTION Found throughout America north of Mexico, wherever it is sufficiently damp for them to survive.

GARDEN ACTIVITIES Millipedes are generally detritivores, although some feed on living plants. Centipedes are carnivores. Symphylans are detritivores, but may feed on roots.

Spinning and Weaving Spiders

Spider silk has many uses besides the familiar webs used to capture prey. Silk is used for wrapping prey, making egg sacs, nests, and drop lines. Some spiders even leave a silken trail by which they can find their way home.

Orb webs, the familiar symmetrical round webs that catch flying insects on sticky silk, are one of the first things that come to mind when we think of spiders in the garden. These relatively sedentary snare builders do a lot of work for us in our gardens even as they remain motionless for hours, heads facing down, in their webs. Studies show that spiders eat more insects than do birds and bats combined!

Despite their relatively small size, spiders loom large in the folklore and myths of many cultures. Arachnida, the name of the taxa that includes spiders, stems from the Greek weaver Arachne, who challenged the goddess Athena to a contest and was subsequently turned into a spider. The Navajo learned to weave blankets from a girl who was taught by Spider Woman. In West Africa, we find stories of the trickster Anansi (tales that later traveled with the slave trade to the Caribbean). The British band The Who related the tale of an ill-fated arachnid in "Boris the Spider." We sing to our children about itsy-bitsy spiders climbing up the waterspouts, while impressing upon them a fear of hairy, scary spiders.

When the large black and yellow garden spider, *Argiope*, appears in fall gardens, we are surprised and wonder where she came from. Of course, these

Diadem spider (*Araneus diadematus*)

spiders have been present all along, but as fall approaches, females increase dramatically in size. They make prominent zigzag shapes resembling letters like X, V, and W in their webs. The shape, called the stabilimentum, may reflect UV light to lure in flying insects; research is ongoing. The only non-venomous spider family, the Uloboridae, also creates webs with stabilimentums.

Not all web-building spiders are as easily spotted or identified. Many species are masters of camouflage and mimicry. Some species resemble twigs, grasses, plant buds, or bird droppings. Some spiders like to hide in damp, dark corners, a reason they are often featured in B-level horror movies.

Orbs are not the only kind of prey-catching spider web. Most of us have seen funnel-shaped webs in low vegetation. The spiky, fast-moving grass spider sits at the entrance, running out to capture prey and drag it back to the shelter of its hideaway. These spiders like to connect their funnels to an existing retreat of some kind. In my garden, a grass spider once built its web inside a bamboo stake near my front door, placed at a height where I could easily observe the goings-on.

The mesh or tangle-web weavers make irregular meshes (cobwebs or tangle webs), often with strands reaching down to the ground. The extreme stickiness of the silk made by these comb-footed spiders allows them to successfully capture prey many times their size. While you may not spot cobwebs in your garden because they are often in hidden or out-of-the-way locations, some other webs may escape your notice simply because they are so tiny. Sheet-web weavers (of those horizontal hammocklike webs so easily spotted when grass is damp with dew in the morning) are also extremely abundant and diverse. Very important in agricultural ecosystems, these include diminutive dwarf spiders, with their correspondingly small webs, and more.

Among the garden residents that spiders help control are familiar nemeses such as aphids, armyworms, leafhoppers, spider mites, thrips, plant bugs, cucumber beetles, grasshoppers, scarabs, and flies.

Spiders cannot control what flies into their webs, although many reject distasteful prey by cutting unwanted corpses loose. This means that some beneficial insects will get trapped from time to time. But webs can act as selective filters. Some insects are able to avoid webs; for instance, hover flies are rarely caught. And the cobwebs and sheet-webs, being low to the ground or in the leaf litter, are best suited to capturing hopping and walking insects and catching plant-sucking insects.

Fortunately, it is easy to integrate spider habitat into your garden. Flowers attract flying insects for an abundant food supply and the long, sturdy stems of plants with decorative seed heads act as scaffolds for webs. Keep in mind that silk is composed of an unusual protein that uses a lot of energy to produce (spiders conserve by consuming their old web before constructing a new one); give them a hand by leaving their webs undisturbed when practical. Hot coffee in hand to brace against the autumn chill of early morning, I love to venture into the garden to marvel at the sparkling dew-spangled webs, strung like jewelry throughout the fall stems of salvias and grasses.

Black and yellow garden spider (*Argiope aurantia*)

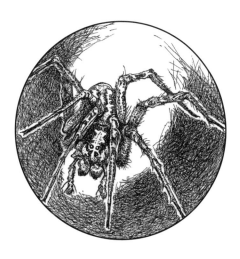

Funnel-web spider (Agelenidae) in her funnel

APPEARANCE Eggs always in a silken cocoon (spheres, teardrops, discs) attached to a substrate. Spiders range from virtually microscopic to a body length of 3.5 in. (90 mm) and a spread of 11 in. (280 mm) across the legs.

POPULAR NAMES Orb-weavers, sheet-web weavers, funnel-web weavers, mesh weavers.

SCIENTIFIC NAMES Class Arachnida, order Araneae.

COMMON TAXA Orb-weavers: Araneidae (garden spider), also Uloboridae, Theridio-somatidae, and Tetragnathidae. Funnel weavers: Agelenidae (grass and funnel spiders). Sheet webs: Linyphiidae (dwarf or micro spiders). Meshes: Theridiidae (comb-footed spiders, e.g. black widow) and Pholcidae (daddy-long-legs). Some spider species use a combination of web types.

DISTRIBUTION North America, more than 2090 species. Worldwide, over 44,000 species in over 128 families.

GARDEN ACTIVITIES Predaceous on insects, mainly flies; but also beetles, grasshoppers, and butterflies. Sometimes prey on other arthropods or rarely, small vertebrates.

Wandering and Hunting Spiders

Spiders are venomous, but very few have fangs that can penetrate human skin. Fangs and venom are used to subdue prey, not humans! Did you know that spiders eat many species of disease-spreading flies in addition to other insects?

Although all spiders make silk, many do not use it to construct orb webs or other kinds of snares for prey capture. Different kinds of silk are produced by several different glands (spinnerets) in the spider's body, and employed for a variety of uses during all their life stages: nests, egg sacks, and "draglines" that anchor spiders safely to a substrate at intervals. Spiderlings (baby spiders) of some species—as well as some adults—extrude a silk strand on a warm day until caught by a passing breeze, whereupon they float away to a new home. They may be transported vast distances in this manner, called "ballooning." No doubt some came to your garden that way.

Regardless of how they got there, many of the spiders in your garden make a living by hunting and pouncing. Some of the most common are wolf spiders, jumping spiders, and crab spiders. Most are active at night, but many of these wandering spiders are active at dusk or even in daylight. All three are easy to recognize, and all are harmless garden friends.

Wolf spiders are frequently encountered garden denizens. Many live in burrows and may dash out of mulch and groundcovers when you water or a sprinkler's spray hits the area. The females often have an egg case attached to their spinnerets, or you may see newly emerged spiderlings riding about on the

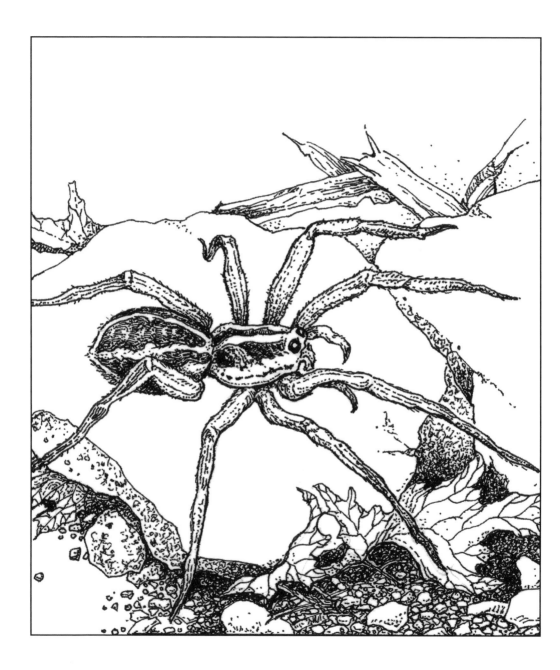

Wolf spider (*Schizocosa mccooki*)

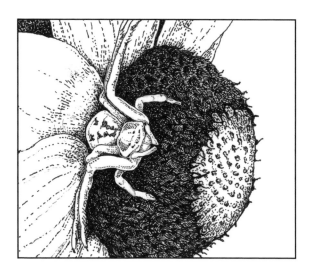

Crab spider (*Misumenoides formosipes*)

back of the mother's abdomen. Despite their name, like almost all spiders, wolf spiders are solitary, and don't hunt in packs.

Jumping spiders are vigorous hunters, stalking prey and pouncing once in range. Their eyesight is extremely acute, and they quickly scurry to the other side of a stalk or leaf when too closely watched by an inquisitive gardener. The two large central eyes give them a look of curiosity when peering back at us.

Crab spiders are ambush hunters. They are most often noticed motionless on blossoms, where they prey on visiting insects. Often pale in color, they hold their legs out to the side. Named for both their appearance and their ability to move dexterously sideways, crab spiders are often beautifully camouflaged, blending with their floral background. They do prey on pollinators and other flower visitors, but are not so abundant as to reduce the population.

Like web spinners, hunting spiders are carnivorous, generalist predators, but unlike passive webs that trap whatever flies into them, these hunters' methods let them pick and choose their prey. While crab spiders sometimes eat pollinators, including bees, most spiders tend to avoid stinging insects as well as those that are distasteful to them. Instead they often focus on a wide variety of flies, small beetles, and true bugs.

Generalist predators have been thought to be less effective in suppressing pest populations than specialists, but recent studies have shown that this is not always the case. Spiders in particular are critical to regulation of insect populations in many habitats. Spiders also provide an important food source for other invertebrates, as well as reptiles, amphibians, birds, and the occasional mammal.

Encourage these allies in your garden with plantings of groundcovers and perennial bunch grasses, which provide valuable habitat. Leaf litter is an important overwintering resource. Planting shrubs and perennials, leaving dead flowers alone instead of removing them all ("deadheading"), and including flowers that attract insect prey are all good strategies in a spider-friendly garden.

Despite the considerable contribution spiders make, it can be difficult to overcome the fear. While spiders (with very few exceptions) have poison glands, their venom serves primarily for paralyzing prey and only secondarily as a defense against predators. In the majority of cases, their venom has little effect on us mammals, and most spider bites are no worse than a wasp sting. Spiders are generally timid, and most won't bite even when handled (the great majority cannot even penetrate human skin). The few North American spiders with venom known to harm humans, such as the brown recluse and black widow, are very seldom found in gardens and, like almost all spiders, are shy, nonaggressive, and easily avoided.

Jumping spider (*Phidippus johnsoni*)

APPEARANCE Eggs always in a silken cocoon, in various shapes: spheres, teardrops, discs attached to a substrate. Juveniles are a smaller version of the adult form. Spiders can be virtually microscopic; our largest North American spiders are the tarantulas.

POPULAR NAMES Wandering, hunting, ambushing spiders. Wolf spider, crab spider, jumping spider, and many others.

SCIENTIFIC NAMES Class Arachnida, order Araneae.

COMMON TAXA Wolf spider (Lycosidae), crab spider (Thomisidae), jumping spider (Salticidae).

DISTRIBUTION North America, about 1460 species, plus more than 140 tarantulas and trapdoor spiders. Worldwide, about 35,000 species in over 100 families.

GARDEN ACTIVITIES Carnivorous, dining on insects, mainly flies, and sometimes millipedes, sowbugs, and even other spiders.

Snails and Slugs

Are snails and slugs actually sluggish? Well, yes. The top speed of the banana slug, a large and relatively speedy West Coast species often found in redwood forests, is only 6 inches per minute.

I know what you are thinking—snails and slugs are garden *allies*?! Now you have gone too far! Okay, I won't pretend that I have not had my share of cursing the slithery and the slimy. As someone who loves to grow salad greens, I won't argue the point. It would take a real shift in perspective to welcome them to the garden. I will simply say that they are indeed an important part of the food web. Many reptiles, amphibians, arthropods, and birds rely on them for food. Let's get to know them better, and I will share some techniques and tricks for controlling them in your garden.

It was purportedly a Frenchman, Antoine Delmas, who introduced the common brown garden snail to California in the 1850s, because he missed eating escargot. (Can we forgive him if I also tell you that he introduced an obscure grape varietal that now produces California's renowned zinfandel?) Shall we respond to an incursion by grabbing a sauté pan, butter, and garlic? Most gardeners probably seek alternatives. However, this brings up an important point. Most problem snails and slugs are non-native, and they have proliferated in the absence of natural enemies.

By the way, a slug is not a snail without a shell. They are an entirely different species. A snail's shell is an integral part of its body; its organs extend from

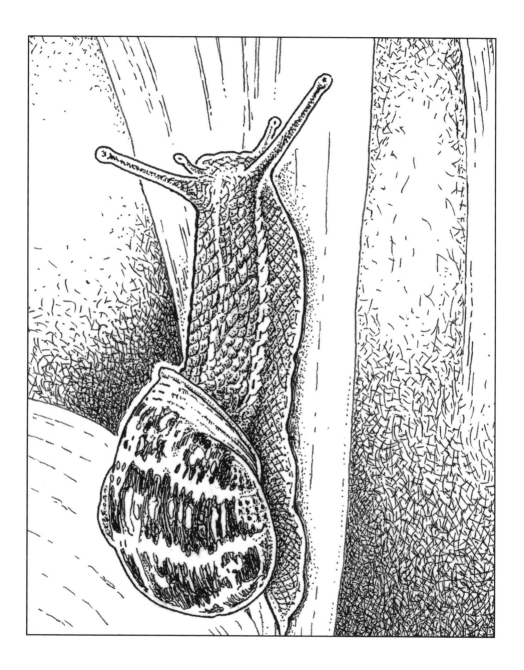

Brown garden snail (*Cornu aspersum*)

its body into its shell and it will perish without it. Slugs, on the other hand, have no shell (although there do exist semi-slugs that may have a tiny one). Snails are able to completely retreat into their shells. In hot summers, they can estivate (hibernate) by sealing the opening to their shell with a parchmentlike membrane (the epiphragm) that helps conserve moisture.

In both snails and slugs, the top of each of their upper tentacles has an eye-spot that can only detect light. The two bottom tentacles are for feeling and smelling. In temperate climates, most species are hermaphrodites; each individual has reproductive organs of both sexes, and can lay eggs after mating. Adults deposit clusters of pearly eggs in the soil. Learn to recognize those when you see them and dispose of them. Garden snails usually live for a year (one full growing season, in colder areas), though some can survive up to three years. Slugs, on the other hand, only live for six to eighteen months.

Snails and slugs are land-going molluscs (class Gastropoda, the "stomach-foots"). As that name implies, they get around on a muscular organ called a foot. They need moisture, as they continually excrete mucus and travel on a slimy trail. They have a radula, a sharp sort of file, used to eat vegetation in your garden. They love succulent, tender foliage and flowers. Hostas, salad greens, strawberries, and garden flowers (lilies, especially) are on their menu. If you go out with a flashlight in the evening, especially after a rain, you may be surprised to see how many snails and slugs live in your garden.

One of the simplest as well as most effective ways to combat snails and slugs is to take a bucket along with that flashlight. Collect all you see, including the tiniest juveniles. If you or a neighbor keep chickens, give them to the birds in the morning; they will be thrilled. Or, turn them into wild bird food. I know a gardener who puts slugs on a rock island in the middle of her birdbath. If they try to escape, they drown. If they stay put, birds swoop down and dine on them. Otherwise fill that bucket with soapy water and drown them. If you do this on a nightly basis for a week or two, the population will be substantially reduced.

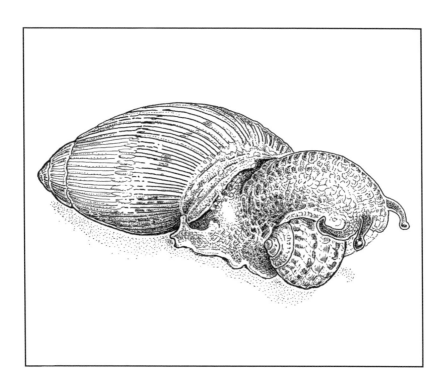

Predatory snail (*Euglandina rosea*)

It is also easy to trap snails and slugs. Set out some upside-down empty grapefruit or melon halves, old cabbage leaves, or other substantial tender vegetation, and collect and dispose of your quarry in the light of day. Or put out pie tins of cheap flat beer; they really do climb in and drown. You can then toss their carcasses in the compost pile or bury them.

Copper barriers set up around the bases of prized plants you want to shield are an effective if more costly solution. The barriers must be wide enough and positioned carefully. If any part of a plant is hanging down, the pest just uses this as a bridge over the obstacle. The salts and acids in slug slime have an adverse biochemical reaction when they come into contact with copper; the barrier essentially electrocutes them.

Baits of any kind should be used judiciously and in a targeted way, as they kill all gastropods, native or not. They are best used in conjunction with other methods. Diatomaceous earth is effective, but needs to be reapplied after each watering or rainfall. My preferred baits are the Sluggo products, as they are non-toxic to other animals. These granules contain iron phosphate; scatter them on (damp) ground around plants you wish to protect, ideally in the evening when the slugs and snails are most active. It suppresses their appetite when they eat it, and they crawl off and die within the week. Bits they miss will degrade and become a harmless part of your garden soil.

A remedy you might not know about is predatory snails. There are two species of native predatory snails that are touted as controls for their unwanted relatives—the science is still unclear on their effectiveness. Be glad if you have them in your garden, but don't introduce them, as they will eat native as well as introduced species. Some species of marsh flies (sciomyzids) are snail and slug predators; their potential is being investigated for control of the gray garden slug.

I once made a casual aside to a group of biologists at a backyard barbeque that our chickens could recognize the sound of snails being tossed into their pen for snacks (as opposed to rocks). The chickens would race over to nab the tasty morsel (unlike us, they don't require garlic and butter). I was immediately peppered with questions from my scientifically inquisitive friends: what was the sound of rocks versus snails landing, what was the size of the snails, the size of the rocks, the density of each, the reaction of the chickens to other materials. I still have not run any experiments, and those hens are long gone. But recently we started keeping a small flock again, and there are still snails to be controlled. I have a small pile of rocks at the ready.

APPEARANCE Common slugs and snails tend to blend into the background; they are usually shades of brown or gray. Spiral shells of snails almost always wind clockwise; a few species and even families' shells coil counterclockwise (sinistrally).

POPULAR NAMES Snails and slugs.

SCIENTIFIC NAMES Phylum Mollusca, class Gastropoda. Phylum Mollusca includes class Gastropoda (snails and slugs); others are marine animals. Gastropoda includes all snails and slugs from saltwater, freshwater, and land.

COMMON TAXA In California, brown garden snail (*Cornu aspersum*); in Florida, the predatory snail (*Euglandina rosea*); elsewhere, gray garden slug (*Deroceras reticulatum*).

DISTRIBUTION Most Mollusca are marine and herbivorous (cephalopods are mostly predators) and widely distributed; there are over 80,000 described extant species; about 80 percent of these are snails. Snails and slugs are found in any garden, but species vary depending on geographic location.

GARDEN ACTIVITIES In gardens, they feed on living and decaying plants. Carnivorous species prey on other gastropods.

Not What They Appear to Be

White oak trees sometimes host jumping galls—the tiny "detachable" galls drop to the ground by the tens of thousands, where the larva encased within cause them to bounce about in a frenzy.

The often strange-looking growths sometimes spotted on twigs, stems, and leaves are not a normal part of a plant. Instead they are induced by insects or other invaders, including fungi, viruses, bacteria, mites, nematodes, and even mistletoe. The word "gall" comes from the Latin *galla*, meaning "oak apple," a conspicuous large gall caused by a diminutive and rarely noticed chalcid wasp. Oak apples are a common sight on oak trees, including native valley oak trees throughout the West and on red and white oaks in the rest of North America. While oaks are the most common host, other plants, including willows and wild roses, can have a great diversity of galls.

Gall-inducing insects are generally highly host-specific. Hundreds of species of cynipid wasps, for instance, are associated with the various oaks, a particularly rich source of study for "cecidologists," scientists who investigate galls. Cecidologists also point to other species of wasps, flies, and aphids as agents of gall formation in North America. In Australia, psyllids and thrips may also be responsible.

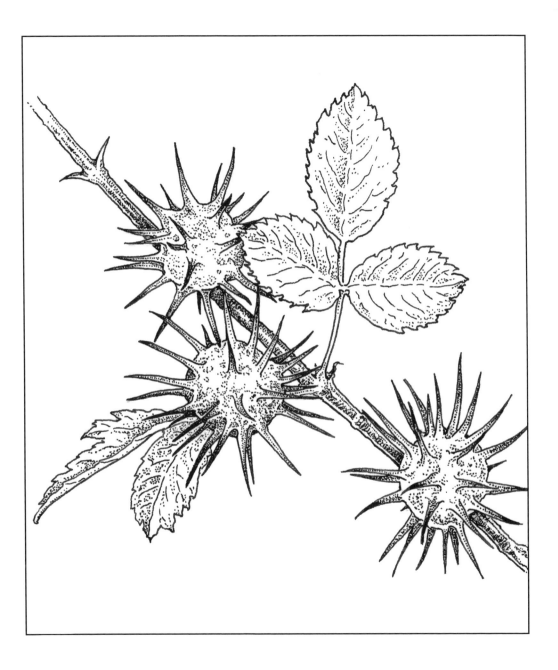

Spiny bud gall on rose

Galls take many forms, some of them highly ornamented. What accounts for all this variety? The answer is chemistry. It turns out that chemical signals are excreted by the ovipositing insect or by larval saliva. These cue the plant to produce a particular size, shape, and color of gall. And what fantastical forms they take! As a result, as you would expect, it often becomes possible to identify the initiating insect by identifying the gall.

A gall often begins when a wasp or other insect deposits its egg or eggs in plant tissues. The plant then responds by walling off the invader with a swelling, or tumorlike, growth. The gall provides both food and a relatively safe environment for the larva within. Galls are generally annual, though many stay on plants long after the insect or other inducing organism has vacated.

Most galls do not harm or kill a host plant. But sometimes damage is caused by a secondary agent. An acorn gall wasp, for example, creates a pathway for bacteria to enter acorns, causing an abundance of sticky froth, unpleasant to walk on, and sometimes leading to weakened trees.

Some galls are pests, for instance canker gall, which can damage orchard trees, and the galls caused by the balsam woolly adelgid on true firs. Over the centuries, phylloxera has brought repeated episodes of devastation to the wine-grape industry, both in this country and abroad. Phylloxera nymphs induce protective galls on the undersides of grapevine leaves, weakening and eventually killing the vines.

However, galls are occasionally a great benefit to gardeners. A classic example is the root nodules caused by nitrogen-fixing bacteria such as *Rhizobium* and the actinomycetes species in the genus *Frankia*. There are more details about these in Chapter 1.

Gall wasps can even be pollinators. Figs, long in cultivation in the Middle East, have a critical relationship with a certain chalcid wasp. Each fig species has its own species of wasp associated with it. Fig flowers are internal in twigs (an uncommon, specialized structure called a synconium); the pollinating wasp must enter one end, mate, and lay eggs before moving on and transferring some pollen to the next synconium. Back in the 1880s, when people began trying to raise figs as a crop in California, no fruit set because the right wasps were not present. The ultimate solution, however, was not to import and use the correct wasp, which works but proved cumbersome. Instead, parthenocarpic figs (those that do not require pollinators to develop fruit) were developed. These are the majority of figs we grow today.

Galls serve other useful functions. In the food web, many provide important food to birds that seek out the larvae within. Some are used for making dyes and for tanning. Ink from galls was noted for its long-lasting quality, and was used by monks in Europe, in the signing of many historic treaties, and even by the United States Treasury for printing currency. Galls have been used for medicinal purposes for centuries, for everything from eyewash, to treatments for cuts and burns, and even as remedies for hair loss.

Galls form quickly in spring and early summer, when plant growth is most rapid. Try collecting a few fresh galls and

California oak gall

Manzanita leaf gall

placing them in a jar, and watch for the insects that caused them to eventually emerge. You may be surprised at the variety of insects that appear. Why would this happen? One reason may be that oak apples and some other galls, once mature, are often occupied by insects behaving as inquilines, causing no harm to the original occupant. The interlopers are merely finding shelter in a cozy home (and having an occasional nibble). Gall inquilines sometimes kill another insect if it should cross its path. A gall wasp's larva may also be parasitized by another species of wasp, and sometimes the parasite will in turn be parasitized. Up to five levels of "hyperparasitism" can occur, echoing mathematician Augustus De Morgan's immortal remark: "Great fleas have little fleas upon their backs to bite 'em, and little fleas have lesser fleas, and so ad infinitum."

Gall-inducing agents can also lead to plant deformities; these do not kill the host but can mar their appearance. Witches' brooms, those dense bundles of shoots that arise from a common point sometimes seen in trees, are one example. Another is fan-shaped fasciation as is sometimes seen on certain cacti and other plants. With their many intriguing stories, galls can provide a lifetime of study for the amateur naturalist or professional cecidologist. Aren't we lucky that our gardens can offer us such a rich diversity of galls to engage us in exploring nature?

APPEARANCE Galls take on many unusual and fantastical forms related to the plant on which they are found and the agent causing the gall. Size varies.

POPULAR NAMES Oak apples, willow shoot galls, witch's broom.

SCIENTIFIC NAMES Many species of Cynipidae wasps induce oak galls.

DISTRIBUTION Widely distributed in America north of Mexico; the West Coast has a particularly high diversity.

Pathogens in the Garden

Bacillus thuringiensis (*Bt*) is deadly to caterpillars (and only to them). Caterpillars may be vegetable garden pests that must go, or they may be those that metamorphose into the beautiful butterflies floating about our gardens. Pathogens should always be used strategically.

Raising caterpillars in an enclosed environment until they emerge as butterflies is a delightful project. By protecting caterpillars from the predators and parasitoids that normally reduce their numbers, a far greater success rate than would occur in the wild can be achieved. That is, until the day the caretaker makes the gruesome discovery of limp, black caterpillars hanging from the leaves! The culprit, in this case, is a pathogen. Sometimes the pathogen is on the leaves you are feeding to the caterpillars, and they ingest the spores, but it could also be airborne.

Pathogens are defined as disease-producing microorganisms that slow growth, reduce reproductive success, shorten life, and/or kill their host. They are one of the principal mechanisms that regulate insect populations, and under the right conditions can even dramatically reduce insect population numbers (an event called epizootic). I sometimes reflect that it is remarkable that any insects at all survive to maturity, given how they are beset from all sides by all manner of threats, not only from predators, parasites, and parasitoids, but also from bacteria, viruses, fungi, nematodes, protozoans, and other disease-causing organisms! These many threats are the reason why insects lay so many eggs.

To the observant gardener, some of these diseases have curious symptoms. Infected insects may change in color, or appear swollen. The limp, black

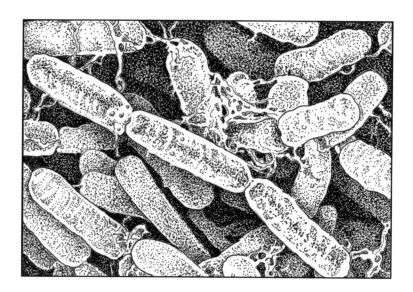

Bacillus thuringiensis colony

caterpillars discovered in our butterfly enclosure were attacked by a virus that left them liquefied and ready to burst and spread the infection onto the plant to which they were still attached. Plants that would be munched upon by the next generation of caterpillars, perpetuating the problem. Bacteria, fungi, and protozoans all can attack caterpillars, each with its own set of symptoms and different manner of infection. When a pathogen strikes, sorry to say, there is nothing to be done but sanitize the enclosure and start over.

Fungal diseases that attack insects can have some strange, disturbing-looking but interesting results. A BBC special a few years ago explored this phenomenon in a piece dramatically titled "The Zombie Fungus and the Climbing Dead." Afflicted insects climb to the top of a blade of grass, leaf, or twig before they die, continuing to cling even in death. As they perish, the spore-filled fruiting bodies of the attacking fungi emerge to create a furry coat. The net result is that fungus has been elevated to a suitable height for dispersing its spores.

Less commonly seen, though found the world over, are many species of *Cordyceps* fungi. These have some intriguing properties. They parasitize various insect larvae, including caterpillars, in the process producing a mushroom, usually bursting out of the head of the by-now dead victim. One caterpillar-attacking species of *Cordyceps*, commonly known as "Himalayan Viagra," has an active ingredient that is mostly used to treat erectile dysfunction. These fungi also have been used in Chinese and Tibetan medicine for centuries, and have been a focus of research in Western medicine. Lab studies show that some species can stimulate immune cells in humans (a possible cancer treatment?), and perhaps also improve kidney function.

Fortunately, most insect pathogens are perfectly harmless to humans and other mammals. Furthermore, they are usually host-specific, sometimes restricted to a single insect species. They may even be specific to only certain life stages. While many insect pathogens are the subject of interest for research scientists looking for benign ways to control insect pests, only a few of these are commercially available and even fewer are practical for home gardeners. Most useful pathogens' restricted host range and slow-acting nature, combined with the difficulties of producing a commercial product, will continue to cause limitations.

At least two bioinsecticides are widely available, however, and can be useful to gardeners. The now well-known bacterium *Bacillus thuringiensis* (*Bt*) is highly effective at killing unwanted caterpillars, such as those that plague crops like broccoli and cauliflower in our vegetable patches. Use it with caution, though, as it kills all caterpillars, including those of monarch butterflies, swallowtails, and other desirable species. My advice is to carefully target the pests you wish to kill, and only deploy it when there is no wind to spread it around.

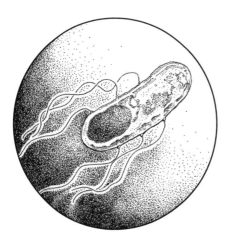

Bacillus thuringiensis individual

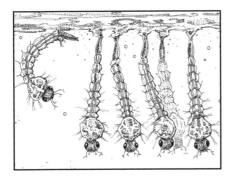

Mosquito larvae

Bacillus thuringiensis subspecies *israelensis* (*Bti*) can be used to control mosquito larvae in backyard water features. If you have a thriving garden pool, however, proceed with caution. Inert ingredients in commercial formulations of *Bti* are mildly toxic to *Daphnia*, a tiny crustacean that is an important component of aquatic food webs. The essay on dragonflies and other aquatic insects in Chapter 5 has more details.

A few other biopesticides are already in widespread use. A nematode has been deployed in Florida to control mole crickets. Some fungi are available in the agricultural battle against grasshoppers and against household cockroaches. A virus has been used in the battle against the devastating gypsy moth. While not suitable for home gardens or commercial crops, a protozoan is in use as a long-term control of grasshoppers in rangeland. In all cases, effectiveness may rely on environmental conditions such as temperature and moisture, or on the abundance of the target pest. Although it is too soon to recommend most of these to home gardeners, there is great promise for the future. Stay tuned!

Avenues for future exploration include antagonists, pathogens that attack injurious plant pathogens. In some cases, fungi, which attack a variety of hosts and can spread quickly among insects, are carried by parasitoid insects. They may be particularly effective for controlling sucking insects, such as white-flies, aphids, and leafhoppers, which are unaffected by pathogens that must be ingested. In the meantime, I will be exploring recommended techniques for keeping pathogens out of my butterfly enclosure, such as ramping up my sanitation efforts and collecting butterfly eggs before the caterpillars emerge.

APPEARANCE Most go unnoticed by the casual observer; sometimes, it is the absence of the insect the pathogen is attacking that is noticed, rather than the pathogen itself.

POPULAR NAMES *Bt* and *Bti*, other bacteria, nematodes, fungi, protozoa, viruses.

SCIENTIFIC NAMES *Bacillus thuringiensis* and *Bacillus thuringiensis* subspecies *israelensis*, *Cordyceps* spp.

DISTRIBUTION Many pathogens are present in the environment; some are produced and sold for agricultural, commercial, and residential use.

GARDEN ACTIVITIES Attacking insect pests.

7

HIGH AND LOW

Birds, Bats & Other Vertebrates

We have explored the lives of many small, even microscopic allies in these pages. Of course, many of the insects, myriapods, and others are certainly large enough to spot easily. Vertebrates, however, while present in fewer numbers, are larger and thus can have a greater impact on our gardens per individual (especially the larger herbivores like deer). Here we will focus on smaller vertebrates that can more easily be integrated into garden ecosystems.

An inspirational garden tour opened my eyes to the value of vertebrates. One property owner's emphasis was on homes for toads, frogs, lizards, and birds and she had created the most remarkable habitat. The garden was alive with activity! As I entered, I noticed a series of flue tiles sunk sideways into the hillside next to a gravelly path. Nearly all of these had a resident lizard sitting on its stoop. Songbirds flitted around the flowering plants, singing melodiously. I could hear but not see some chirping frogs in the vicinity of a small pond. I was enchanted.

Another garden on that same tour, beautifully landscaped, contained a surprising element. A huge snag (a tall dead tree), smack in the middle of it, had been spared the saw. Gazing up at it, we could all see it provided habitat for various nesting birds, in addition to a family of acorn woodpeckers who were stashing their food supply in holes they'd drilled into the trunk. If you have a similar situation and the dead tree doesn't pose any safety hazard, I hope you'll consider doing the same.

Many vertebrates are desirable in our gardens, in particular many amphibians, reptiles, birds, and bats. Most eat insects during at least part of their lives. Many flowering perennials and shrubs, especially native species, support the caterpillars, bees, wasps, and flies they rely on for food. The arrival of the vertebrates inevitably follows—as the saying goes, if you build it, they will come.

A habitat garden that welcomes vertebrates does require some extra thought. Frogs are drawn to a water feature, but many vertebrates benefit from the shelter of diverse plantings as well as brush piles, rocks, even logs in quiet corners.

In particular, I would like to put in a good word for garter snakes. They are harmless, and eat insects, slugs, and even some larger garden prey such as mice. They are likely to want to avoid any human contact. Your garden may have already hosted such a visitor or have a few in residence, and you may not even know it. As for bats, the more insects, the better!

Admittedly, any conversation about attracting birds is sure to produce a lively debate among gardeners. My pea seedlings have more than once been pulled from the ground, and I have lost my share of ripe berries to birds. But overall, they add an essential, beneficial, and aesthetic element to any garden. While most birds eat a varied (omnivorous) diet, almost all terrestrial bird species in North America eat insects for at least part of their lives. Many birds seek out nutritious insects only during the breeding (including egg-laying and baby-feeding) season, when protein becomes a necessity. Not coincidentally, breeding season for birds, late spring and summer, is exactly when many insects and other arthropods are also abundant. Thorny plants (for nesting), fruiting and seed-bearing plants, and conifers and deciduous trees all offer valuable resources for birds. Choose ornamental berried plants with care, however; invasive species are often spread by berry-eating birds. Native plants are a better choice. Note that the Audubon Society strongly recommends that domestic cats be kept indoors.

Birds can be categorized by their diet and method of obtaining food. Some glean insects and other arthropods directly from plant surfaces or from the ground. Other birds catch insect prey in flight. Raptors help keep the rodent population under control. Not all birds fit neatly into a single category—nature is rarely that simple. Orioles glean from leaves, but also catch insects on the wing. Hummingbirds snatch fruit flies in midair, in addition to gleaning, and they don't perch or give aerial chase, but hover.

Many creatures are part of the big picture, not only of your garden's health but of the wider ecosystem. When you welcome and encourage them, you'll have less and less need for pest "remedies" (from insecticides to baits and traps)—and gardening will become easier, more fun, and ultimately more rewarding.

Toads, Frogs & Salamanders

Because amphibians breathe largely through their skin in addition to their lungs, they must stay wet to absorb oxygen. Juveniles such as tadpoles and some adult amphibians breathe through gills.

Frogs, toads, and their relatives the salamanders are not the first creatures that come to mind when we are thinking about garden allies, but they are a great addition to the team. One useful characteristic unites these amphibians: generalist predators, they eat heaps of insects, grubs, slugs, and other invertebrates!

Frogs and toads are easily distinguished. Watch their gait—frogs have longer legs and jump, while toads have shorter legs and hop or walk. Also, frogs have smooth, mucus-covered skin, and toads have rough, bumpy, thick skin. Recognize salamanders by their elongated body, lizard-shaped but free of scales (they are not even closely related to lizards). Their skin is moist and fairly smooth. Both toads and salamanders are equipped with glands that excrete toxins that protect them from predators, a reason to avoid handling them.

Somebody gave me a toad once, which I released into a small outdoor lath house we called our "jungle room." At times, swinging gently in the hammock, listening to the mesmerizing sounds of a "Tropical Rainforest" recording, I would spy Mr. Toad emerging from his home. It was my belief that he was

Pacific chorus frog (*Pseudacris regilla*)

California slender salamander (*Batrachoseps attenuatus*)

lured out by the recorded deluge. Toads can live for many years in one place, and my family was sad when our warty pet finally departed.

Here in California, wherever there is fresh water, we have Pacific chorus frogs; back East and in the Midwest, the dominant species is the related spring peepers. When I visit them in their natural habitat, no matter how stealthily I approach, the raucous chorus inevitably pauses. If I wait quietly and patiently, one voice eventually braves the silence, then another, until finally the full cacophony resumes. Despite their loud voices, they are very small and hard to see due to coloration that blends in with their surroundings. They are also nearly impossible to catch.

As tempting as it may be, gardeners should refrain from catching wild amphibians to bring home. Many species are threatened, plus there are regulations on the capture and transportation of most species. Raising tadpoles is still a wonderful project to undertake with a child, where it is allowed. Please check local regulations, and return the small froglets to their wild home once they develop their legs.

The better, wiser tactic is to lure amphibians to your garden. Assuming you don't live too far from a nearby pond that's already populated with them, you may be able to coax native amphibians to your garden. Bear in mind that frogs and toads begin life in water (jelly-coated eggs are laid in water in bundles or strands) but become terrestrial as adults. Some salamanders also lay eggs this way; others bypass the aquatic phase entirely. Unfortunately, our old cast-iron tub, sunk into a garden corner, does not harbor our delightful little native frogs. That's because it is frequented by predatory dragonfly larvae, which relentlessly hunt down any other larvae (including tadpoles). If frogs were my priority, I would empty the tub annually, thus eliminating those marauding dragonfly larvae, while also discouraging bullfrogs (invasive where I live), which eat other frogs, but whose larvae require a full year to develop.

It is possible to entice adult frogs and toads into your garden in other ways. Small piles of rocks, logs, or boards in secluded areas can offer shelter. Make a "toad abode" from an overturned clay pot; prop it up on a rock or chip a little entrance. Set it in the heart of an area that is troubled by slugs, earwigs, or other unwanted garden interlopers. Remember that shade is important—toads don't like getting overheated. Mulch well around it, and insects and other arthropods will move in, offering good and easy meals for the toad. Last but not least, please remember: don't use garden pesticides and herbicides around frogs and toads, as they are particularly sensitive to poisons.

Even in urban areas, gardeners with woodsy gardens may be pleasantly surprised to find a salamander residing under a pot or rotting branch. A gardening friend, sitting on her patio in a residential neighborhood one evening, was startled to spot a giant Pacific salamander appear from behind a pot. Her home is near a large wetland, the probable origin of her visitor. This formidable salamander can attain 12 inches, among the largest land salamanders in the world. If handled roughly, it may bark or snarl, and can deliver a painful bite! In other parts of the country, harmless little yellow-spotted, dark brown salamanders (3 to 8 inches long) become active in wet spring weather, their mating season. They spend the rest of the year mostly underground, in rocks or caves or other

Eastern American toad (*Anaxyrus americanus*)

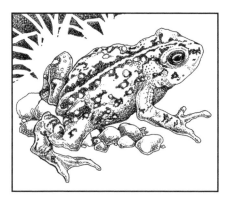

California subspecies of western toad (*Bufo boreas*)

nooks, dining on arthropods they find in the soil.

Gardeners can play a role in the protection and preservation of these creatures. Many are threatened in the wild, for multiple reasons, not the least of which is habitat loss. As noted above, they require water at the beginning of their lives and the destruction of wetlands and vernal pools eliminates important breeding grounds. Many species migrate and if watery habitat is no longer available, or if they have to cross a busy road, their prospects dim. Perhaps your garden can become a small sanctuary.

Amphibians serve as a biological indicator of environmental change, often reacting to increases in pollution and UV radiation by exhibiting distressing mutations. Many species appear to be particularly sensitive to climate change, and cannot survive even small changes in temperature. In addition, there's the problem of invasive species—other species of amphibians as well as fish and crayfish that prey on native amphibians. If that weren't enough, a fungus (chytrid) that attacks delicate amphibian skin, is the culprit in some regional extinction events.

Good reasons, all, to create amphibian-friendly habitat in our gardens.

APPEARANCE Amphibians have four legs and glandular, moist skin with no scales, feathers, or hair. Coloring is highly variable, sometimes even within one species.

POPULAR NAMES Frogs, toads, salamanders, newts.

SCIENTIFIC NAMES Class Amphibia; two orders in gardens: Anura (frogs and toads) and Urodela (salamanders).

COMMON TAXA Pacific chorus frog (*Pseudacris regilla*) and spring peepers (*Pseudacris crucifer*). Bullfrog (*Lithobates catesbeianus*), native to eastern United States and Canada, invasive in other regions. Eastern American toad (*Anaxyrus americanus*) and the western toad (*Bufo boreas*). On the West Coast, slender salamander (*Batrachoseps* spp.); on the East Coast, spotted salamander (*Ambystoma maculatum*).

DISTRIBUTION North America, over 300 species; 11 families of frogs and toads; 9 families of salamanders.

GARDEN ACTIVITIES Amphibians eat insects, slugs, earthworms, sowbugs, and other invertebrates.

Snakes, Lizards & Skinks

Alligator lizards maintain a higher body temperature than most other lizards, allowing them to be more active in cooler weather and to prey on other reptile species.

Other than spiders, perhaps no other animals in the garden engender as much fear and misunderstanding as reptiles. Once safety has been ascertained, however, the curious gardener may draw near to appreciate these scaly allies who keep many of our arthropod populations under control. Garden reptiles consume insects, spiders, slugs, and even larger garden prey such as rodents.

Reptiles commonly found in gardens include snakes and, in warmer climates, lizards (including skinks). If you are lucky, you may even attract native turtles. Snakes tend to lie low, camouflaged by their surroundings, while lizards may dart around, especially when preferred food is abundant. Among the places you can find any of these are rock walls, drain pipes, culverts, and mounds of twigs and soil. If you provide such shelters for them, as with the flue tiles I saw in the garden I described earlier, position them so the creature is safely out of the way of foot traffic that would disturb or startle them.

We used to think of reptiles and amphibians as cold-blooded, but they are more properly called ectotherms, deriving their body temperature from the environment rather than internal processes. It turns out that reptile body temperatures can vary, and some are able to maintain even higher temperatures than some warm-blooded animals (endotherms).

Common garter snake (*Thamnophis sirtalis*)

Gopher snake (*Pituophis catenifer*)

Reptiles are long-lived, certainly compared to most of the smaller creatures we have spent most of this book exploring. Garter snakes may live up to five years, while skinks can live up to twenty years. All reproduce by laying eggs (with the exception of some snakes). Should you come across any eggs during garden maintenance, please leave them undisturbed.

Garter snakes (*Thamnophis* spp.) are by far the most common garden resident snake in North America, especially in gardens that are near streams or swampy areas. Excellent garden allies, they mostly eat insects. Depending on where they are found, they may exhibit a variety of coloring, but are almost always striped longitudinally, a useful camouflage for slithering through grass.

Also sometimes found in our gardens are the closely related ringneck snakes (*Diadophis punctatus*). In addition to other reptiles, they eat slugs and earthworms. A few years ago, I came across a couple as I cleaned up a long-abandoned corner of a new garden. Although these small and graceful snakes are harmless, when disturbed they coil up to reveal a bright orange underbelly and may emit a fetid chemical.

Rattlesnakes are prevalent in the western states. Their diet consists of rodents and, sometimes, small birds. Should you find one in your garden, I recommend that you call a reptile rescue organization for a relocation service, rather than killing it or scaring it into some unknown spot. Gopher snakes (*Pituophis catenifer*) are more widely distributed. While they can be mistaken for rattlesnakes, they are distinguished by the lack of rattles and a slimmer head. The gopher snake's habit of flattening its head and vibrating its tail when threatened emphasizes the similarities, however, and has caused many a frightened gardener to kill this beneficial snake. As the name implies, it will eat gophers, as well as other rodents it may encounter in your landscape.

Constrictors such as gopher snakes and pine snakes (*Pituophis* spp.) eat rodents, birds, and frogs. King snakes (*Lampropeltis* spp.) principally prey on other reptiles, although they also eat other animals. Several color variations exist, depending on region. Interestingly, they are immune to the venom of rattlesnakes; I once watched in awe as a king snake ate a young rattler.

Western skink (*Plestiodon skiltonianus*)

Western fence lizard (*Sceloporus occidentalis*)

Lizards, on the other hand, are less intimidating to most gardeners than snakes, and welcome in the mild-climate areas they favor. You'll occasionally find these insect-eaters in a woodpile or heap of rocks. In the same forgotten corner where I found those ringneck snakes, I also spotted the neon-blue flash of the tail of a young western skink (*Plestiodon skiltonianus*) as it escaped into the safety of the leaf litter. Skinks are distinguished from other lizards by their short limbs and thick necks, which lend them a snakelike appearance. As western skinks mature, their tails lose the beautiful strikingly azure blue color. Like other garden lizards, skinks eat a variety of insects, roly-polys, snails, and other invertebrates, and can detach (autotomize) their tails, which then thrash about and presumably distract hungry predators.

Coming across a full-grown alligator lizard (*Elgaria* spp.) is startling. These can be over 7 inches long, not including the tail. Able to be active at lower temperatures than most other lizards, they are effective predators of skinks and other lizards, in addition to spiders, snails, and insects such as crickets and grasshoppers. Although they're active during the daylight hours, they are rarely seen basking in the sun like other lizard species, and in hot spells, may take cover during midday.

Anolis is a huge genus, and though most are not native, in much of Florida, the South, and along the Gulf Coast, the native American green anole (*Anolis caroliniensis*) is very common. Usually greenish or brown, they can change color depending on heat and light. They range in size from 3 to as much as 9 inches and are elusive and fast-moving. Gardeners in these regions may appreciate them for their voracious appetites. They consume insects but also eat just about anything smaller than themselves, from insects to spiders and various sorts of worms.

The western fence lizard (*Sceloporus occidentalis*) is a common lizard in the West, with a bright blue belly, often seen basking on rocks in the sun. Males may occasionally be seen doing "pushups" to impress nearby competing males.

Their diet is insects and spiders. Interestingly, they have a protein in their blood that kills the Lyme disease bacterium carried by ticks; consequently, the incidence of this disease is lower in regions where these lizards are found. Leaving my home on summer mornings, I observed one fellow that always seemed to be waiting expectantly in the same spot. The mystery was solved when I learned that a neighbor enjoyed hand-feeding resident lizards with pet-store mealworms. What a pleasure to have wild insect-eating pets!

APPEARANCE Always scaly. Snakes are legless. Skinks have the head broadly joined to the body and small legs.

POPULAR NAMES Snakes, lizards, skinks.

SCIENTIFIC NAMES Class Reptilia includes the order Squamata (lizards and snakes), discussed here. Reptilia also includes the testudines (turtles), and the Archosauria, which comprises Crocodilia and Aves (the birds) due to their close relationship. Reptiles usually refer to non-avian Reptilia.

COMMON TAXA Gopher snake (*Pituophis catenifer*), pine snake (*Pituophis melanoleucus*), garter snakes (*Thamnophis* spp.), king snakes (*Lampropeltis* spp.), and ringneck snakes (*Diadophis* spp.) Alligator lizards (*Elgaria multicarinata* and *E. coerulea*), western fence lizard (*Sceloporus occidentalis*). Three genera of skinks: *Eumeces*, *Neoseps*, and *Scincella*.

DISTRIBUTION *Pituophis* and *Thamnophis* snake species and skinks are found throughout America north of Mexico; alligator lizards are on the Pacific Coast; western fence lizard is found in the western United States. Anole lizards are found throughout the South, in Florida, and along the Gulf Coast.

GARDEN ACTIVITIES Depending on taxa, they eat arthropods, small invertebrates such as earthworms and slugs, amphibians, other snakes, and a variety of birds and rodents.

The Gleaners

Woodpeckers are equipped with a long, extendable tongue for probing insect galleries. Their tongue may be three times the length of their bill, wrapping around the skull when not in use.

Standing at my kitchen sink one day, I watched a hummingbird hovering alongside a spider web that stretched between an old rose and the nearby windowsill. I knew that hummingbirds use spider silk to line their nests, but it soon became clear that this bird was poaching insects from the web, using it as a convenient "pantry." That day, I learned that hummingbirds, well known as nectar-feeders, also eat insects, and they can often be observed gleaning aphids and other small arthropods from plants.

Gleaners usually refer to birds that pluck insects and other arthropods from leaves or crevices. Here we also consider birds that feed on the ground or drill into trees. Gleaners show preferences for searching specific parts of plants. Warblers and bushtits, for instance, pluck insects from foliage. Titmice, chickadees, and nuthatches harvest insects from crevices in tree bark as well as from leaves. Woodpeckers, such as flickers, probe the trunks and main branches of trees. Towhees find insects in leaf litter on the ground. Warblers, tanagers, and grosbeaks feed on insects in foliage in the high canopy of trees, whereas sparrows, jays, and some other ground-gleaners venture into low-growing vegetation in search of an insect meal. However, many will make an exception for a tempting morsel and occasionally give chase on the wing.

Anna's hummingbird (*Calypte anna*)

Some—notably wrens, vireos, nuthatches, and brown creepers—are almost exclusively insectivores. Most insectivorous birds eat a variety of insects, but some have more specific tastes. Orioles, for instance, love to eat wasps, potential good news for those of us who like to eat outdoors on a back deck or patio. Warblers have a varied insect diet and employ multiple feeding methods. Jays are omnivorous, but insects are a significant component of their diet, especially during nesting season. Juncos are primarily seed eaters, but also feed insects to their young. And while we always think of robins wresting worms out of the garden ground, they also eat a lot of insects.

Many herbivorous birds relish an occasional non-plant treat. The goldfinches that cling to swaying stems of cosmos in fall also consume insects as they pluck off seeds. Quail eat insects while they look for the leaves and seeds they prefer. Pheasants eat everything—grasses, roots, seeds, and berries, as well as grasshoppers and crickets, beetles, ants, caterpillars, not to mention earthworms, snails, and slugs.

Although some gleaning birds that eat insects are migratory, most are resident. Creating year-round habitat is worthwhile in order to sustain as many different species as you can, for as long as possible. Hold back on removing flower heads ("deadheading") in your borders in fall, allowing the birds to feast on flower and grass seeds (and the insects that eat those seeds). Bird favorites include millet and many members of the aster family: coneflowers, sunflowers, cosmos, coreopsis, and black-eyed Susan. Attractive berry- and fruit-bearing plants are easy to include in your garden; there are many, but top choices include chokeberry, crabapple, cedar, serviceberry, hawthorn, viburnum, and winterberry; and, in the West, native toyon, snowberry, and wild currant species. Remember to check fruit-bearing plants for invasiveness in your region when choosing landscape plants. Many birds also enjoy pecking around tufts of bunch grasses for food.

In addition to offering appealing food sources, offer birds additional reasons to linger in your yard. Sound is an often an overlooked garden element; dripping or bubbling water turns out to be attractive to birds. Birds appreciate a shallow birdbath for drinking as well as bathing; be sure to keep the water topped off and clean. Provide secure footing. Positioning a bird bath where birds will feel safe can take a bit of experimentation. Nearby branches for perching, and a clear view to safety, are the minimum requirements. A mix of tall plants and shorter ones gives birds the "ladder" they favor to assess safety as they descend, while also offering shelter and places to build nests. Depending on the species, birds require different nesting zones and types of shelter; thorny shrubs will be especially appreciated by nesting birds. Quail, a ground-nester and forager, is attracted to brush piles and brambles.

To get started, find out which birds are likely to live in your neighborhood. Then maintain or add plants that will provide suitable food and habitat. Perhaps erect a birdhouse, to entice specific species. Remember to leave windows open to allow birdsongs from the garden into your home! A beautiful and diversified garden will be satisfying to both human and avian inhabitants.

Bushtit (*Psaltriparus minimus*)

Red-shafted flicker (*Colaptes auratus*)

APPEARANCE Winged and feathered with two feet. Often have distinctive markings. Many insectivorous birds are quite small, and continually active as they search for food.

POPULAR NAMES Gleaners include quail, hummingbirds, woodpeckers, flickers, vireos, jays, titmice, chickadees, bushtits, nuthatches, creepers, wrens, robins, wood-warblers, old-world warblers, grosbeaks, tanagers, juncos, towhees, sparrows, orioles, finches, and old-world sparrows.

SCIENTIFIC NAMES Perching birds or songbirds—passerines (order Passeriformes)—include over half of all bird species. Quail are an upland species (order Galliformes).

COMMON TAXA Most discussed here are common in gardens. Orioles, tanagers, woodpeckers, and flickers are less common but are not rare.

DISTRIBUTION Many garden birds have a wide distribution; some are seasonal.

GARDEN ACTIVITIES Insectivorous for all or part of life cycle. Gleaners sometimes feed arthropods to their nestlings, but may feed primarily on seeds or foliage as adults.

Feeding on the Wing

Cliff swallows sometimes practice a type of brood parasitism, laying eggs in other females' nests of their own species. Occasionally, they may even carry their eggs from their own nest to another's nest.

One of the pleasures of the college campus near my home (California's Sonoma State University) is a colony of cliff swallows that nests high on the walls of one of the buildings. Watching their graceful flight is always a welcome diversion, and I looked forward to their return each spring. Cliff swallows, along with barn swallows and phoebes, have adapted well to the intrusion of humankind, which provides convenient nesting sites on buildings, bridges, and under the eaves of houses. And what a fortuitous thing that is! Barn swallows wheeling and swooping and twittering above my garden in the evening hours are always a delight, not only as entertainment, but also because mosquitoes are one of their favorite foods.

Aerial hunters are aces at catching insects on the wing. Some birds, such as orioles, are primarily gleaners, but will go after flying insects when the opportunity presents itself. True aerial insectivores, such as swallows, hunt on the move, often swooping back and forth through swarms of insects to catch their prey. Nighthawks and nightjars are aerial insectivores that hunt nocturnal insects. Nighthawks are also well-adapted to disturbed habitat, and thrive in urban environments, where they can be found nesting on flat rooftops. Swifts are perhaps the most aerial insectivores of all, spending most of their lives and

Black phoebe (*Sayornis nigricans*)

carrying out most of their feeding activities on the wing. Other birds are known as hawking hunters (or salliers); from a comfortable perch, they scan the vicinity for food and sally forth when prey is spotted. A few, such as western bluebirds, feed from perches close to the ground.

Most such birds are fairly indiscriminate in their choice of insects, but some have definite preferences. Swifts search out insects such as swarms of flying termites and ants. Swallows feast on mosquitoes, and several owl species eat moths and other large nocturnal insects. Flycatchers pursue many species of insects—especially flies, earning their name. Some flycatchers, such as phoebes, either catch their prey near the ground or even snatch it off the ground, but quickly take flight again. Warblers and cedar waxwings sometimes catch insects in flight.

Attracting aerial insectivores to your garden is more complicated than with the more familiar garden birds. They are always on the go, and may not find suitable nesting spots on your property. Try providing a nesting box or platform; note that different species have different requirements. Some—for example,

White-throated swift (*Aeronautes saxatalis*)

Pacific-slope flycatcher (*Empidonax difficilis*)

purple martins—need a fully enclosed box with an entry of exacting diameter, installed with a specific exposure, and at a minimum height. Other birds are satisfied with a simple platform tucked under an eave; black phoebes can often be found nesting in such situations. Barn swallows, as their name indicates, are frequently found in old barns and other large buildings in rural areas. Detailed information is easily found in bird books and on birding websites. Short of that, offer perching spots on wires or clotheslines to encourage these birds to linger.

Open space, ample room for wheeling, swooping, and diving is important. A broad lawn or natural meadow, front or back, may draw these birds. Let your grass grow a bit higher, which encourages more insects for them to feed on as well as a good area over which to skim.

Water, too, is essential. Because aerial hunters tend to drink by skimming over the surface rather than perching to sip, a bigger garden pond or even a stream could attract some of these birds. On the other hand, any garden water source may be of interest to thirsty birds. They sometimes show up and hurtle through the broad spray of lawn sprinklers. Don't be surprised when new bird species arrive as time passes and your garden matures.

My friends will tell you that the plants in my garden have been chosen to attract insects. There is no doubt, however, that the plants I have included also attract insectivorous birds. As I watch the activity, I am sometimes led to think that my garden is really for the birds!

APPEARANCE Winged and feathered with two feet. Often have distinctive markings. Many insectivorous birds are quite small, and continually active as they search for food.

POPULAR NAMES Nighthawks, nightjars, swallows, shrikes, swifts, bluebirds, phoebes, flycatchers, waxwings, warblers, and more.

SCIENTIFIC NAMES Most aerial hunters are passerines, order Passeriformes. Nighthawks and nightjars are in order Caprimulgiformes; swifts are in order Apodiformes.

COMMON TAXA Nightjars and nighthawks may be less common than some other aerial insectivores; swallows and warblers are more commonly seen.

DISTRIBUTION Many common garden birds have a wide distribution; some may be only seasonal visitors.

GARDEN ACTIVITIES Many aerial hunters are completely insectivorous; some eat a widely varied diet, or eat insects only occasionally. While some birds are indiscriminate in choice of arthropods, others specialize on certain insect orders. Aerial insectivores in all stages of life feed on insects.

Talons for Hunting

Red-tailed hawks can spot a mouse from 90 feet! Swooping down, they scoop up their victim in their talons, and fly back to their perch for a leisurely meal.

Standing in our narrow side yard with my husband one evening, we were startled by the sudden arrival of a red-tailed hawk, which scooped up a gopher snake less than 15 feet away. It seemed at first to be a rare experience, but a quick informal survey of a few gardening friends revealed that many have a story about a close encounter with a backyard raptor. Sharing curved talons for capturing prey, and hooked beaks for tearing, raptors vary greatly in size, general appearance, and life histories. They are generally well camouflaged and can be hard to spot unless they're moving. Like their prey, we tend to be unaware of them until they strike!

Among the raptors that hunt where humans live are the true falcons (in the genus *Falco*, which includes peregrine falcons and American kestrels) and hawks (including ospreys), a general term that refers to all raptors other than vultures, falcons, and owls. In addition to talons and sharp, hooked beaks, raptors have superior eyesight and, in the case of owls, excellent hearing.

These larger birds tend to have long lifespans. While some owl species may live only a year or two, some hawks can live ten to fifteen years. They are also creatures of habit, many raptors traveling great distances from wintering grounds to spring breeding grounds and returning to the same nesting site

American kestrel (*Falco sparverius*)

Red-tailed hawk (*Buteo jamaicensis*)

Barn owl (*Tyto alba*)

year after year. Large raptors hatch one or two young annually.

Many raptors are shy of human contact and have suffered declining populations due to habitat destruction. Some are still recovering from the devastating effects of DDT, which caused eggshells to be so thin that they would break long before hatching. However, Cooper's hawks, peregrine falcons, and western screech owls are well-adapted to human habitats, and sometimes nest in urban and suburban environments, where prey can be plentiful. In the wild, raptors favor a variety of habitats, from riversides to cliffs to open ground, in both coniferous and deciduous forests. Most require trees or other tall objects for nesting. Their nests may be hidden in tree crowns, on ledges, and in cavities. Many prefer specific species of trees for nests. Others don't build nests at all, instead using existing hollows or other birds' abandoned nests.

Raptors have evolved to fill similar ecological niches. They may avoid competition with one another by hunting at different times of day—for example, owls are most active after dark. Their body forms reflect hunting strategies and prey preference. Agile red-tailed hawks are ambush hunters, perching and waiting to sight prey. Barn owls rely on hearing; the facial disc they possess focuses sound on their sensitive asymmetrical ears.

As for diet, some raptors are highly specialized, relying on a single or only a few prey species for survival (for instance, some kite species eat only snails, and osprey eat fish). White-tailed kites eat primarily mice and voles, while sharp-shinned hawks and Cooper's hawks prey mostly on songbirds. But most have a broader diet that allows them to survive more easily when prey is scarce. Red-shouldered and red-tailed hawks' main prey is rodents, but they will eat many other small animals. American kestrels eat insects, including large drag-onflies, grasshoppers, and crickets, and may take small birds. Peregrine falcons mainly catch birds on the wing. Barn owls primarily hunt rodents and western screech-owls hunt many small animals, including insects, while great horned owls eat a broad diet, but mostly mammals.

Any of these may make an unannounced visit in your garden and relieve you of a rodent, bird, or reptile. One yard is unlikely to be sufficient for long, and you don't need to find ways attract them. One exception: if your property is plagued by gophers, try installing a nesting box for barn owls (one family of barn owls may result in the control of over 3000 rodents in a single season!). For full details, find owl expert Steve Simmons's useful article "Build a Barn Owl Box" on the internet.

What a thrill it is to hear the arrival each spring of the red-tailed hawks that nest in one neighbor's towering eucalyptus tree, and to hear the resonant hoot of the great horned owls that nest on another neighbor's property. Personally, I plan to add a barn owl box to our garden. It will be a pleasure to listen for the sounds of a nesting pair.

APPEARANCE Vary greatly in size, color, and general appearance; all have talons and hooked beaks for tearing up their catch; owls have forward-facing eyes. In most bird families, males are larger than females; the reverse is true of raptors.

POPULAR NAMES Falcons, kites, osprey, harriers, accipiters, buteos, and owls.

SCIENTIFIC NAMES Class Aves. All raptors listed above are order Falconiformes, except owls, order Strigiformes.

COMMON TAXA Infrequent garden visitors; most common are red-tailed, red-shouldered, and sharp-shinned hawks, but kestrels and owls may also visit.

DISTRIBUTION About 30 species of Falconiformes and 19 species of owls in North America. Many widely distributed, some only seasonal.

GARDEN ACTIVITIES Carnivores. They eat rodents and other mammals, reptiles, amphibians, fish, insects, birds, and even other raptors. A few raptors are specialists; most are fairly indiscriminate.

A Pitch for Bats

Consider establishing a bat colony with a bat box, installed high on a building. Most bats are insectivores. Did you know that a single tiny pipistrelle bat can eat up to 3000 insects per night?

Wherever they are found, bats play a vital role. Most North American bats are exclusively insectivores, preying on many agricultural pests in addition to mosquitoes. Three flower-visiting species are key pollinators in our desert regions. In tropical climates, bats are also critical to seed dispersal. If you go out in the evening hours and watch bats' energetic flights, you can be reassured that they are scooping up and ingesting lots of insects in your neighborhood. A single little brown bat (*Myotis lucifugus*) can catch up to 1200 mosquito-sized insects in an hour!

Bats are the only true flying mammals. If you were to stretch out a bat's wing, the resemblance to our human hands would be instantly apparent. While bats are often thought of as flying mice, they are not in the order Rodentia but rather in their own mammalian order, Chiroptera. Unlike rodents, most bear only one or two young per year, and a typical lifespan is about 20 years, although some species live only a few years.

Like other mammals, bats live and hunt with their own kind. Some species have different roosts for giving birth and for overwintering, and sometimes, different species may share roosts. True, like other mammals, bats can contract rabies, but the danger of getting the disease from them is greatly exaggerated; we are in more danger of dying from a dog attack or bee sting. Just don't attempt to handle them.

Little brown bat (*Myotis lucifugus*)

Yet fear of bats persists, and is reflected in language such as "like a bat out of hell," "batty," and "bats in the belfry," and in their associations with vampires and rabies. Incidentally, bats are not "blind as a bat," but generally possess excellent eyesight. Most insectivorous bats, however, find their prey through a remarkable feature called echolocation, emitting high-frequency sounds (not discernible to human hearing) that bounce off nearby objects and back to the bat's ears to pinpoint the location of any nearby object. Consider this: if bats are able to detect objects as small as a mosquito, they certainly aren't going to fly into your hair, a common fear.

Several bat species can be spotted in gardens. The western pipistrelle, the smallest bat in the United States, is common in desert regions and lowlands, and is often the first bat to appear in the evening in those areas. The comparatively large pallid bat can be seen flying close to the ground and, unlike most bats, takes its insect prey from the ground. But the ones that are the most widely distributed and most often seen are the high-flying common little brown bat and the big brown bat.

The latter two are among the species susceptible to white-nose syndrome, a fungus that has devastated bat colonies in North America. Another cause of bat decline is overuse of pesticides for mosquito control and on crops. These farm chemicals reduce the insect populations on which they feed, contaminate the insects that survive, and thus poison the hunting bats. Wind turbines and habitat loss due to human activity have also adversely affected bats. Sad to say, they are now the most endangered warm-blooded animals in North America; over 30 percent of North American bat species are imperiled. If these worrying trends continue, our ecosystems will certainly suffer.

Gardeners can do our part to help, however. First of all, do not use pesticides, so that the bats will have plentiful and safe food to eat. The preferred roosts of the bats mentioned here include barns and houses. While we want to keep our home interiors free of bats, why not put up a "bat house" (buy

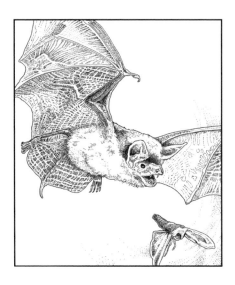

Mexican free-tailed bat (*Tadarida brasiliensis*)

one, or make your own) outside and invite these insect-dining mammals to take up residence? For best results, hang the box up high, 10 to 15 feet above garden level, tacked either to a sturdy pole or post or attached to a building. Another trick is to switch on a bright incandescent flood-light in your yard after nightfall. In short order, you'll see moths, mosquitoes, and other flying insects in the beam, and soon the bats will arrive to feed. Personally, I am quite content to simply take a twilight walk in my neighborhood and watch the bats emerge as the dusk deepens.

APPEARANCE North American species vary from the tiny western pipistrelle, with a 7- to 9-in. (180–230 mm) wingspan, to the hoary bat, with a 15-in. (380 mm) wingspan. Coloration of bat species varies dramatically, from reddish, grey, yellowish, to pale brown, and combinations of these.

POPULAR NAMES Big and little brown bats, pallid bats, Mexican free-tailed bats, western pipistrelles, hoary bats.

SCIENTIFIC NAMES Class Mammalia, order Chiroptera. North American bats are in suborder Microchiroptera, distributed among four families. Vespertilionidae includes most bats discussed here, with one exception: the free-tailed bats, family Molossidae.

COMMON TAXA Big brown bats (*Eptesicus fuscus*) and little brown bats (*Myotis lucifugus*) are the most commonly seen. Also, pallid bat (*Antrozous pallidus*), Mexican free-tailed bat (*Tadarida brasiliensis*), western pipistrelle (*Parastrellus hesperus*), and the common but rarely seen hoary bat (*Lasiurus cinereus*).

DISTRIBUTION 47 species in North America; more than 1400 species worldwide.

GARDEN ACTIVITIES Bats in North America are almost exclusively insectivorous; three species are nectar and pollen feeders.

EPILOGUE

GARDENING MATTERS

The first gardens were utilitarian, cultivated for food, medicine, fiber, and other human needs. But appreciation for gardens' ornamental value was surely not far behind. Throughout much of garden history, pollination was taken for granted, with plant-eating insects and other animals viewed as competitors. The first enclosed gardens were undoubtedly meant to keep out large, hungry herbivorous animals, and over the years we found creative and sometimes highly toxic ways to battle insects. Put into practice what you have learned here about conservation biological control—the more natural enemies you have in your garden, the better pest populations will be regulated. The fewer pests, the less you may be tempted to reach for a solution. After a while, you may wonder why you ever used pesticides.

It's high time to find new ways to garden and to contribute to the long-term sustainability of our human-managed landscapes. As our population grows, and concerns about pollinators and environmental health grow, strictly ornamental gardens kept pest-free with chemicals are increasingly out of favor and indeed out of step with the times. In addition to regulating populations of plant-eating insects, you can support pollinators, promote biodiversity, nutrient cycling, and soil conservation by practicing the methods outlined in this book. The way you approach the design of your garden also affects hydrological processes and erosion control. Far from being "a drop in the bucket," your efforts truly matter, as gardens cover many acres of land across North America.

We can do this! In fact, the myriad ways I have shown you to manage your garden's allies are fairly easily implemented in a diverse garden setting. Your garden, your community, and the planet will thank you.

Gardener's essential tools

METRIC CONVERSIONS

INCH	MM
⅛	3.2
⅙	4.2
¼	6.4
⅓	8.5
⅜	9.5
½	12
¾	19
1	25

TEMPERATURES

$$°C = \tfrac{5}{9} \times (°F - 32)$$

$$°F = (\tfrac{9}{5} \times °C) + 32$$

CM	CM OR INCHES	INCHES
2.5	1	0.4
5	2	0.8
8	3	1.2
10	4	1.6
13	5	2.0
15	6	2.4
18	7	2.8
20	8	3.1
23	9	3.5
25	10	4
51	20	8
76	30	12
100	40	16
130	50	20
150	60	24

CM	CM OR INCHES	INCHES
180	70	28
200	80	32
230	90	35
250	100	40

M	M OR FEET	FEET
0.3	1	3.3
0.6	2	6.6
0.9	3	9.8
1.2	4	13
1.5	5	16
1.8	6	20
2.1	7	23
2.4	8	26
2.7	9	30
3	10	33
6	20	66
9	30	98
12	40	130
15	50	160
18	60	200

M	M OR FEET	FEET
21	70	230
24	80	260
27	90	300
30	100	330

GLOSSARY

ACTINOMYCETE Bacteria; mostly in soil and decomposing organic material; some fix nitrogen.

ACULEATE Stinging hymenopterans; includes bees, ants, some wasps.

APID Member of the Apidae bee family. Includes honey, bumble, carpenter bees, others.

APOSEMATIC Coloration warning predators of danger. Venomous, stinging insects and mimics.

ARACHNID Arthropod class includes spiders, harvestmen, scorpions, ticks, mites, other forms.

ARTHROPOD Invertebrate with exoskeleton, segmented body, jointed paired appendages.

AUGMENTATIVE BIOLOGICAL CONTROL Intentional introduction of native or non-native arthropod for control of pest arthropods.

BACILLUS THURINGIENSIS (BT) Bacteria used for control of lepidopteran (butterflies, moths) larvae.

BACILLUS THURINGIENSIS SUBSPECIES ISRAELENSIS (BTI) Bacteria used for control of mosquito larvae.

BEETLE BANKS/BUMPS Raised earthen berm; habitat for ground beetles, spiders, other predatory arthropods.

BIOINSECTICIDES Living agents used to control pest populations; arthropods, fungi, pathogens and more.

BIOLOGICAL CONTROL Control of pests using predators, parasites, and herbivores.

BLUE-GREEN ALGAE Cyanobacteria. Single-celled alga; some fix nitrogen for plant use.

BRACONID Member of Braconidae wasp family; often parasitic on pests.

BROOD PARASITES Egg laying in nest of another individual. Social insects and birds.

CARNIVORE/CARNIVOROUS Meat-eating animal.

CATERPILLAR Lepidopteran (butterfly and moth) larva.

CLASSICAL BIOLOGICAL CONTROL Intentional introduction of non-native arthropod for control of (usually) non-native pest.

CENTIPEDE Predatory myriapod. Class Chilopoda.

CHALCID Member of superfamily Chalcidoidea; includes many parasitoid species.

CHELICERATE Major subdivision of arthropods; includes arachnids and sea spiders.

CHRYSALIS Lepidopteran (butterfly, moth) pupa.

CHRYSIDID Member of family Chrysididae. Jewel wasp.

COCOON A silken covering enclosing a pupa (of Lepidoptera, Hymenoptera, some other arthropods).

CONSERVATION BIOLOGICAL CONTROL Enhancement of habitat to attract and maintain natural enemies of pest arthropods.

COEVOLUTION/COEVOLVE Reciprocal evolution of two species.

COPROPHAGE Dung-eating animal.

CREPUSCULAR Active only at dawn and dusk.

CRUSTACEAN Major arthropod group; largely aquatic.

CYANOBACTERIA See blue-green algae.

DETRITIVORE Detritus- or waste-eater. Decomposers.

DRONE Male bee.

ECOSYSTEM Community of interacting living organisms and their physical environment.

ECTOPARASITOID Parasitic insect living externally on host; eventually kills host. Usually an insect.

ELYTRA First pair of beetle wings; generally hardened or leathery.

ENDOPARASITOID Parasitic insect living internally in host; eventually kills host. Usually an insect.

ENTOMOLOGY The study of insects; often includes spiders, other arachnids.

EUSOCIAL Truly social; colonial insects with cooperative brood care, overlapping generations, and division of labor.

EXOSKELETON External skeleton; arthropod characteristic.

EXTRAFLORAL NECTARIES (EFN) Nectar-producing glands in non-floral locations on plants.

FOOD WEB Interconnected food chains.

FOREWING Front pair of wings; four-winged insects.

FUNCTIONAL BIODIVERSITY Components of biodiversity influencing operation and resilience of an ecosystem.

FUNGIVORE Fungus-eating animal.

GALL Abnormal plant growth in response to injury by arthropods; other agents.

GUILD Group of organisms that exploit the same resources.

HALTERES In Diptera; second pair of wings reduced to knobs.

HEMIMETABOLOUS Incomplete metamorphosis: egg, nymph, adult.

HEMIPTERA Order of "true" bugs.

HERBIVORE/HERBIVOROUS Plant-eating animal.

HERMAPHRODITE Organism with both male and female sex organs.

HETEROPTERANS/HETEROPTERAN One suborder of true bugs.

HOLOMETABOLOUS Complete metamorphosis: egg, larva, pupa, adult.

HOMOPTERA Two unrelated taxa of true bugs.

HONEYDEW Sugary excreta of homopteran insects.

HUMUS Organic matter resistant to further decomposition.

HYMENOPTERAN Order of insects; includes wasps, bees, ants.

HYPERPARASITOIDS/HYPERPARASITISM Parasitoids attacked by other parasitoids; up to five levels.

ICHNEUMONID Member of the Ichneumonidae family of parasitoid wasps.

INQUILINE Animal using living space of another.

INSECTIVORE Insect-eating animal.

INSTAR Stage of metamorphosis.

INTRAGUILD PREDATION Organisms preying on others exploiting same food resource.

INVERTEBRATES Animals without backbones.

KLEPTOPARASITE Animal stealing food from others to benefit offspring.

LARVA(E) Immature stage of insects, other arthropods.

LICHEN Fungi in obligate relationship with an alga or cyanobacteria.

MAGGOT Dipteran (fly) larva.

MANDIBLE Mouthpart.

METAMORPHOSIS Transformation from immature to adult in distinctive stages.

MILLIPEDE Herbivorous myriapod (many-footed), class Diplopoda.

MYRIAPOD Subphylum of arthropods. Many-footed includes centipedes, millipedes, symphylans.

NAIAD Aquatic nymph of hemimetabolous insect; larval form may be very different from adult.

NECTAR Sugary fluid with trace minerals.

NEMATODE Phylum Nematoda; roundworms.

NITROGEN-FIXING Organisms that make molecular nitrogen in air available to plants by converting to usable form.

NYMPH(S) Immature form of terrestrial hemi-metabolous insects.

OBLIGATE MUTUALISTIC RELATION-SHIPS Organisms depending on one another or survival.

OMNIVORE/OMNIVOROUS Plant- and animal-eating animal.

OVIPOSIT Egg laying.

OVIPOSITOR Organ used for egg laying. May be modified into stinger.

PARASITE Lives in or on another organism; benefits at the other's expense.

PARASITOID Insect larvae living as parasites; eventually kills host.

PHEROMONE(S) Chemicals excreted by one organism that trigger response in another of same species.

PROBOSCIS In invertebrates, elongated appendage used for feeding.

PUPA(E)/PUPAL/PUPATE Stage of complete metamorphosis; follows larval stage, precedes adult stage.

RAPTORIAL LEGS Hunting legs, front pair, insects such as some true bugs, mantises.

SPHECID Family of wasps; Sphecidae.

TACHINID Fly in family Tachinidae; parasitoids.

TAXON/TAXA General term for levels of biological organization.

TAXONOMIST Scientist who studies biological organization.

TAXONOMY The organization of related biological organisms.

THRIPS Insects in order Thysanoptera; "thrips" is both singular and plural.

THORAX In insects, middle section between head and abdomen; bears legs and wings.

VESPID Family of wasps; includes almost all colonial wasp species.

RECOMMENDED RESOURCES

Please visit the Garden Allies Facebook page for further, more specific resources.
facebook.com/Garden.Allies/

ELECTRONIC RESOURCES

ALLABOUTBIRDS.ORG/GUIDE/ For identifying birds by appearance, behavior.

BIOCONTROL.ENTOMOLOGY.CORNELL.EDU Excellent website with information on natural enemies of pests.

BIRDS.CORNELL.EDU/NESTINGINFO/ For understanding nests and nesting behavior.

BUGGUIDE.NET Online community of naturalists; packed with information on the natural history of insects and other arthropods.

CULTIVATINGPLACE.COM. Many podcast episodes on garden allies and habitat, with host Jennifer Jewell.

INATURALIST An online community; a great resource for insect (and other organisms) identification.

IPM.UCDAVIS.EDU Excellent website with information on biological control.

TOLWEB.ORG Tree of Life project; information about biodiversity and characteristics of life on earth. If you are interested in evolution, essential reading.

WHATSTHATBUG.COM A great site for insect identification.

XERCES.ORG Website of the Xerces Society; indispensable. They provide books and other publications as well as habitat plant lists suitable for different regions of North America.

Beck, Travis. 2013. *Principles of Ecological Landscape Design*. 2nd ed. Washington, DC: Island Press.

Darwin, Charles. 1881 (OP). *The Formation of Vegetable Mould, Through the Action of Worms*. London, UK. (Available in paperback)

Fabre, Jean Henri. 1917–1928 (OP). *The Life of the Insects*. New York, NY. Dodd, Mead, and Company. (Available in paperback, and scanned on gutenberg.org)

Fortey, Richard. 2009. *Dry Storeroom No. 1. The Secret Life of the Natural History Museum*. London: Harper Perennial.

Grimaldi, David, and Michael Engel. 2005. *Evolution of the Insects*. Cambridge, UK. Cambridge University Press.

Grissel, Eric. 2010. *Bees, Wasps, and Ants: the Indispensable Role of Hymenoptera in Gardens*. Portland, OR: Timber Press.

Lutz, Frank E. 1941 (OP). *Lots of Insects: Entomology in a Suburban Garden*. New York, NY: G.P. Putnam's Sons.

McAlister, Erica. 2017. *The Secret Life of Flies*. Richmond Hill, ON, Canada: Firefly Books.

Nardi, James B. 2007. *Life in the Soil: A Guide for Naturalists and Gardeners*. Chicago, IL: University of Chicago Press.

Owens, Jennifer. 2005. *The Ecology of a Garden: The First Fifteen Years*. Cambridge, UK: Cambridge University Press.

Rainer, Thomas, and Claudia West. 2015. *Planting in a Post-Wild World*. Portland, OR: Timber Press.

Reich, Lee. 2000. *Weedless Gardening*. New York, NY: Workman Publishing Company.

Swan, Lester. 1964 (OP). *Beneficial Insects*. 2nd ed. New York, NY: Harper and Row.

Tallamy, Doug. 2007. *Bringing Nature Home*. Portland, OR: Timber Press.

Van Soelen, Philip. 1979 (OP). *Cricket in the Grass and Other Stories*. New York, NY: Atheneum Books.

Walliser, Jennifer. 2015. *Attracting Beneficial Bugs to Your Garden*. Portland, OR: Timber Press.

Wilson, Joseph S., and Olivia Messinger Carril. 2016. *The Bees In Your Backyard: A Guide to North America's Bees*. Princeton, NJ: Princeton University Press.

Note: Books I recommend are my "essential books," among my most valuable resources. I also list books referenced in the text; some of these may be out of print (OP), but are well worth seeking out. Field guides are not listed; the best ones are regional.

FILMS

The Zombie Fungus and the Climbing Dead. BBC.

The Hidden Beauty of Pollination. Louie Schwartzberg. Moving Art.

Wings of Life. Louie Schwartzberg. Disneynature.

Microcosmos. Claude Nuridsany and Maier Pérennou. Miramax Films.

Fantastic Fungi. Louie Schwartzberg. Moving Art.

ACKNOWLEDGMENTS

A book like *Garden Allies* requires a team of people to bring it to fruition. I am but one member of a fantastic team. Thanks are due to some special friends and colleagues: Richard Turner, former editor at *Pacific Horticulture* magazine, who first invited me to write about beneficial insects, and then pressed me to move far beyond the insects to explore birds, soil organisms, and much more. Lorene Forkner picked up the reins when Richard retired, shepherding the Garden Allies series until its conclusion. I deeply appreciated their mentorship and infinite patience.

This book would not be in your hands if not for my friend Saxon Holt. I am deeply grateful for our conversations exploring the nature of sustainable landscapes, and for his wise encouragement.

My manuscript readers contributed much valuable commentary—all without the pleasure of the beautiful illustrations. Thanks to the expertise of Bob Allen and Philip Van Soelen, I gained confidence in my own knowledge. Other readers, knowing little of the topic, pointed out where I lacked clarity. Some made valuable comments on short sections. Thank you. Thanks are due to my brothers Philippe and Xavier Lavoipierre, and my son Sean Duckworth. I am grateful to have a mother, Michèle Donnet, who pays close attention to the proper use of the English language. To my husband, Robert Duckworth, who now knows more about insects than most people; special thanks are due for all that listening, and also for pointing out the cool insects in our garden.

Craig Latker's drawings brought my text to life; I am grateful for his artistic skill and the new drawings he contributed to this book.

Thanks are due to the many subject experts who generously commented on and corrected essays and essential passages for this book. I, a generalist, owe them each a deep debt of gratitude. Any remaining errors are mine. Scientists are listed here in alphabetical order: R.J. Adams, Dr. Christopher Beatty, Dr. Steven Beissinger, Kate Burroughs, Dr. Hall Cushman, Dr. Gordon Frankie, Dr. Nick Geist, Dr. Matt Gimmel, Dr. Chris Grinter, Dr. Martin Hauser, Dr. Sam James, Dr. Matt Kay, Dr. Lynn Kimsey, Dr. James Nardi, Dr. Tim Pearce, Ron Russo, Dr. Catherine Tauber, Dr. Merlin Tuttle, Dr. Alex Wild (who also contributed photos of honey bees as a model for our drawings), Dr. Dieter Wilkin, and Dr. Doug Yanega. Many of the vital statistics on insects are from BugGuide.com; thanks to John VanDyk for allowing me to liberally borrow for this book. Thanks again to Philip Van Soelen.

There have been many influences on me both as a writer and a scientist. My biology professors at Sonoma State University taught me to think like a scientist; I am forever grateful. J.J. Wilson taught me to write the truth, regardless of the facts, and my friend and colleague Mimi Luebbermann encouraged me to always submit my best work. I cherish my conversations with Dino Rossi, whom I knew for too brief a time and who taught me to appreciate every moment I spend in nature.

Finally, thank you to the team at Timber Press. To Tom Fischer, who had faith that Garden Allies was a terrific topic for a book, to Mike Dempsey, for helping me to successfully navigate the details of publication, and to Teri Dunn Chace, who helped make this book a reality.

Craig Latker sends thanks to Deborah Gibson for never-ending encouragement, to Mary Latker and Marilyn Goode for an early start in art and nature, and to Dr. James Wong for all things ophthalmological.

INDEX

Stacy Bloodworth

FRÉDÉRIQUE LAVOIPIERRE is the author of Garden Allies, a ten-year series of articles published in *Pacific Horticulture Magazine*. She most recently worked at the Santa Barbara Botanic Garden, where she served as director of education. She was previously the founding director of the Sonoma State University (SSU) Sustainable Landscape Professional Certificate Program, where she also coordinated the Garden Classroom and Entomology Outreach programs. Frédérique holds a master's degree in biology, with an emphasis on ecological principles of sustainable landscapes. She currently works as a consultant, and serves on the editorial advisory group for the American Public Gardens Association.

CRAIG LATKER is a professional landscape designer in San Francisco, California. He has a degree in landscape architecture from the College of Environmental Design at the University of California, Berkeley. He has more than 30 years' experience designing creative and sustainable landscapes for private residences throughout California. Craig is also a professional illustrator whose work has appeared in publications by the Nature Conservancy, Sunset Books, and the University of California. He spent ten years illustrating the "Garden Allies" series of articles published by Pacific Horticulture Magazine.